Step into the World of Mathematics

Samuli Siltanen

Step into the World of Mathematics

Math Is Beautiful and Belongs to All of Us

Translated by Lauri Snellman

 Springer

Samuli Siltanen
Department of Mathematics & Statistics
University of Helsinki
Helsinki, Finland

Translated by
Lauri Snellman
Espoo, Finland

This work has been published with the financial assistance of FILI – Finnish Literature Exchange.

FILI FINNISH
LITERATURE
EXCHANGE

ISBN 978-3-030-73345-2   ISBN 978-3-030-73343-8   (eBook)
https://doi.org/10.1007/978-3-030-73343-8

This Springer imprint is published by the registered company Springer Nature Switzerland AG.
The registered company address is: Gewerbestrasse 11, 6330 Cham, Switzerland

Mathematics—Our Invisible Friend

"What use will this ever be?" Every maths teacher has heard the question during lessons on logarithms, derivatives, or arcus tangents. Of course, we can cope with our everyday lives without using them. Even when we are enhancing a selfie, driving to a wedding in Hankasalmi, or clicking the cheapest flight to London on the Internet.

But what is happening under the hood? The Instagram filter raises the numerical values of the pixels into the power of one-half, and goes for a fast Fourier transform on the frequency domain to spice up the outlines of our selfie. The navigator knows our position, because the complex corrective terms given by Einstein's theory of relativity have been programmed into its GPS system. The prices of plane tickets change every second depending on how the clicks of the holiday fanatics having an adventure on the website change a statistical formula.

The math that runs our digital everyday lives is hidden into mobile apps and into workplace computer systems. Even the much talked-about AI runs only additions and multiplications and thus roughly emulates the functioning of the network of nerve cells in the brain. All of these mathematical calculations are hidden, and the layman does not know about it—before reading this book!

In my book, I tell you about this hidden mathematics, introduce its uses and developers to you, and describe how new mathematics is developed. I do not require any mathematical skills from you, dear reader. You can view me as a travel writer, who describes the mysterious valleys and strange ways of life in a faraway country. You do not have to be an expert to enjoy my tales praising the landscapes of the world of mathematics.

These three fellows come up time to time in my story: the average, the intermediate model, and optimization.

You already know averages from your school grades and different statistics. That mathematical operation is modest and simple, but it is powerful! The influence of averages can be felt everywhere, just like the Force in Star Wars. Though you do not have to be a Jedi knight to use averages.

Those Finnish readers who have watched Kummeli probably remember *Kari ja Karvattomat*, a dance band. Its most swinging song goes in English: "*Intermediate model guy, intermediate model song, intermediate model face*". The song gets out of me this artistic interpretation: even though the artist does not sell the Helsinki Olympic Stadium full and the song does not compete for the Finnish Copyright Association's Award, the song can be OK and get the partygoers dancing. But what does Kummeli-Kari's song have to do with mathematics?

A *mathematical model* is a mathematical formula that imitates some phenomenon of life. Robert Malthus' rule for predicting the population of the world is an example of them. The Malthusian growth model determines the time for the doubling of the human population. The prediction holds good for decade-long observations, but over 1000 of years it leads to predictions of 17 people over every square metre on Earth, including the seas. Of course, this cannot really happen, so there is something wrong with the model. The predictive accuracy of the Malthusian model becomes very poor in the long run. The same holds true for every other mathematical model too: they are inaccurate but useful within suitably fixed limits. That is why I sometimes call models *intermediate models,* when I want to remind of their limited validity.

Computer modelling is for the most part based on optimization, which aims at finding the best possible circumstances. For example, a bicycle designer may have fixed a maximum weight for the bike and a goal of finding the best shape for its frame. The bike should take the maximum stress, be easy to manufacture, and also look cool to sell a lot. One can imagine a *penalty* that depends on the shape of the frame to aid in the design process. The penalty grows bigger, when a proposed design goes against the design criteria. The mathematical solution to the problem resembles a hiker skating in a mountain range of icebergs and looking for a way to the bottom of the valley.

I will also tell about how we researchers create new mathematics. This might be a surprise to the proud granny I met on a train, whose grandchild "has already read *all of math*". She could have been thinking that all of mathematics is already finished and collected in some leather-bound book, and the brisk kid hoovers it into his fair-haired head at school. Dear grandma, mathematics is a living science and more of it is born every day! Your grandchild might see it as liberating: there is always more nice stuff to learn!

We all need mathematics every day, without knowing it. But you can learn to know it! Welcome to the world of mathematics. It is reliable, surprising, and most beautiful.

What Is in the Book?

I tell about my own background in Chap. 1. I have liked calculating and solving problems since I was a kid. When I grew up, I ended up first as developing medical imaging technologies for industrial companies and then as a maths professor. I see mathematics everywhere.

Chapter 2 is a gentle introduction to the main theme of the book, *mathematical models.* They are digital helpers that make our thinking more effective just like escalators make moving into higher floors more effective. Mathematical models try to imitate some living phenomenon as accurately as possible by using different kinds of calculations. The models devour data about our world, our messages, our pictures, our thermometers, and our choices of music. They edit and supplement the data. They guide our vacuum cleaners, our timetables, and our feelings.

In Chap. 3, I introduce *climate models* and models for gravity waves that are rippling in the universe. We would not know anything about them without mathematical modelling. I also write about how mathematics could help us to solve the big problems facing mankind. Natural science and technology are a most natural field for mathematics, because "the book of nature is written in the language of mathematics", as Galilei said.

Chapter 4 is dedicated to the use of mathematics as an aid to medicine. I use statistical research and imaging as examples. I especially concentrate on my own field of study of X-ray tomography, which gives doctors a very accurate 3-D map of the internal organs of the patient.

In Chap. 5, I move to a more controversial topic. I ask, if mathematics can describe human thoughts, wants, creativity, or even love? There are already signs of this, when an AI plays chess better than a human being and reaches poetic heights when translating literature.

Chapter 6 is dedicated to the idea that everyone has a right to basic mathematical skills and that everyone should have an equal opportunity to work in mathematics.

I tried to write about mathematics in an easy to approach and fun way. If you hit on a passage you find difficult, please move onwards in the text. Easier stuff can be found there.

Contents

1 My Adventures in the World of Mathematics 1
  1.1 Primary School 1
  1.2 High School 3
  1.3 Engineering Studies at HUT 5
  1.4 Thesis and Medical Imaging Research 7
  1.5 As an Industrial Developer 11

2 Examples of Mathematical Models 13
  2.1 How to Model Gender? 14
  2.2 Natural Numbers Model Quantities 16
  2.3 A Model for Prices in an Online Store 19
  2.4 The Flight of a Football on PlayStation 21
  2.5 Bringing Light to a Virtual Room 22
  2.6 How to Build a Mathematical Model 26
  References 36

3 The Big Models of Earth and Space 37
  3.1 Weather Forecasts, Atmospheric Models and the Butterfly Effect 38
  3.2 Modelling Climate Change 45
  3.3 How to Observe the Collisions of Black Holes? 50
  3.4 Mathematics and the Wicked Problems of Mankind 55
  References 58

4   How Mathematics Helps Doctors   59
   4.1   Why Should Children Be Vaccinated?   60
   4.2   The Fat Lottery and Predictions of Weight   65
   4.3   Traditional Tomography   70
   4.4   The New Low-Radiation Tomography   75

5   Is Humanity Too Just Mathematics?   87
   5.1   A Poor Model Produces Poor Results   88
   5.2   The High Points of AI   91
   5.3   Will Computers Replace Artists?   103
   References   106

6   Mathematics Belongs to Everyone   107
   6.1   The Myth of Genius Is Destructive   109
   6.2   The Properties of an Applied Modeller   113
   6.3   A Challenge for Raising Enthusiasm: let's Bring
       Everybody to the World of Mathematics!   116

1

My Adventures in the World of Mathematics

I'll Reveal It Straight Away I'm thoroughly absorbed by mathematics. I feel that I'm a lamb leg that has been marinading for decades. The garlic of the marinade has been replaced with logarithms, and the olive oil with fractions.

In this chapter I'll tell you about my mathematical background, so that you'll get a better idea of the world I'm inviting you into.

1.1 Primary School

As a small boy, I was a natural science nerd living in Töölö. I did electric experiments on the differences between direct and alternating currents and on voltages. A little electric motor from a slide projector went faster on a 9 volt battery than on a 4.5 volt one. Adding a small adjustable resistor or a potentiometer made possible a smooth adjustment of the speed. Great! Replacing the battery with a 230 volt mains current didn't give an even faster speed, but burnt the fuse and broke the engine.

S. Siltanen, *Step into the World of Mathematics*, https://doi.org/10.1007/978-3-030-73343-8_1

1

On the other hand, the engine of the slide projector didn't work at all on batteries. It required a mains current. Did adding a potentiometer to this circuit slow down the rotation? In a way yes, because the fan stopped, burning the fuse and Mum's nerves with it.

I got a chemistry set for Christmas. After I've gone through the tricks in the manual, it was time to improvise. Adding baking soda to vinegar led to nice bubbles, which I tried to stop by putting a bottle stopper on the test tube. I cleaned up broken glass from my room after the explosion, and Mum wasn't happy on this time either.

My other experiments included putting together a rocket boat out of fuel torn from sparkler sticks and dropping rotating "paper landers" from a seventh-floor balcony. The result was a half a metre flame spouting from toy a boat standing still and parachute strings that were tangled by rotation.

Kids and youngsters: be careful with electricity and fire! It was just luck that I didn't hurt myself with my experiments.

My physics and mathematics experiments taught me a lot of things that don't work. My tests in the field of mathematics gave me many experiences of success, and maybe therefore math became my calling.

As a primary school pupil, I found somewhere a method for calculating square roots with a pen and paper. To my great delight, I determined the square root of 2 more accurately than Dad's calculator (that is, with 13 decimals 14,142,135,623,731). Dad worked at Taivallahti primary school, which was my school. He brought old math textbooks home from there. I solved all their exercises out of the sheer joy of calculating. I felt great happiness when I was able to study power and logarithmic functions years before they were taught in class.

1.2   High School

I developed warm feelings towards sines, cosines, tangents, secants and a certain Maija in junior high. It's a pity that I rarely met Maija. I could always meet the trigonometric functions though. They are still important in my life, but not quite as important as my dear wife Airi.

I still feel that the way sines have been treated has been unfair. In a unit circle, the cosine gives the x coordinate and the sine gives the y-coordinate. The sine should come before the *co*-sine, and x is before y in the alphabet. It's the wrong way around!

My interest of physics also grew, when the science magazine Tiede 2000 came out in the 1980s. I'm grateful to the pioneers of particle physics in Finland, who shared and popularized their knowledge in the magazine while doing their research. I was especially fascinated by the fact that there are only four fundamental forces in the universe.

Gravity was the most familiar to me out of all the four. I had often studied it, like next to a gravel ditch in Hyvinkää when I was 9. My plan for the experiment was based on this argument: since ski jumpers can come down smoothly to a downward slope, a small boy gliding like Superman from the top of the hill will make a soft landing to the sand. My experimental study showed this assumption false. Falling to the ground chest first knocked the wind out of me even though I was going downhill. Gasping for breath gave me a concrete idea of the gravitational force that pulls masses together.

I also got a tangible understanding of the *electric force* with the circuits of the Philips EE series and through the motor experiments I've told you about above.

I was annoyed that my junior high teacher wasn't able to fully explain the two other forces, the *weak* and *strong* nuclear force. Dear Raija Winterhalter: it wasn't your fault. Explaining the nuclear forces to a teenager is difficult even to a professional physicist. You were a great teacher!

My enthusiasm for physics led me to contact Helsinki University when I had a short practical work training at school in 1985. I checked out the address for the physics department from the phonebook and went right to the department at Siltavuorenpenger. I marched in to the High Energy Physics Research Institute without thinking more about my choice, because its door had a name that included the word "physics".

Choosing a job for the training with this partially random method worked out great. I got a good reception, and I got to use the department office's brand new Compact Macintosh computer. I drew works of art that were inspired by the humor magazine Pahkasika with the MacPaint program. My own Commodore 64 couldn't do that!

I had a great time at the HEP, and I got to know the exhilarating atmosphere and computers of the academic world. The bubble chamber photos were mysterious: elementary particles had drawn strange tracks in them. I also got to solder silver wires to the particle detectors in the experiment room, which were used as a base for the real research equipment at CERN. Fortunately my electronics hobby had developed my soldering skills.

I collected feedback papers during my job training for my school. A job description for university assistants stuck into my mind: "opening the door for the professor". Today I'm a professor myself, but I open the doors with my own hands. There are no assistants any more.

Although physics was still a strong competitor when I studied at Linnankoski senior high in Porvoo, I was better at mathematics and more interested in it. Chemistry couldn't beat maths either, even though my teacher Peter von Bagh made great experiments in class.

So I decided to apply for mathematics studies. I felt a strong calling for the general mathematics school at Helsinki University, but Dad's recommendation to go to the Helsinki University of Technology in Espoo was even stronger. I checked it out: HUT offered maths too, so I agreed to pick it. The school was called technical physics, but it was possible to specialize in maths after 2 years of common studies.

1.3   Engineering Studies at HUT

My first year at university was a shock. I had passed my courses at school with little work, but I had to study hard for the exams at HUT. Or I should have, because the duties of student life, that is parties, didn't always leave me enough time for that.

During the spring of my freshman year I found myself a suitable style of study. A differential equations exam was coming, and I couldn't turn the lecture notes into my own understanding at all. Fortunately I found an old and dusty book on the topic in the physics department. It helped me realize, how to solve the problems by assembling a suitable solution out of the general solutions of the equations. The exam went well.

I realized two important things. Once you have found some mathematics, it never changes. One should look for information outside the teachers' course material too.

After this turning point, I was able to balance my studies with an active participation in student activities, including parties. The obligatory physics courses were hard, but I was able to concentrate on the wonder of mathematics once I had passed them. Professor Rainer Salomaa, I'm sorry for the stupid question I made on the course Modern physics II: "Is quantum mechanics nonsense?" I already knew that it's not.

The rugged beauty of abstract mathematics drew me into its depths. I sank deeper and deeper into the whirlpool of theoretical structures that are difficult to explain to people who do not know them. Hard work with books that are abstract and difficult to understand brought into view masses of knowledge, where great crystal mountains and celestial cogwheels were attached to each other like pieces of puzzle lying in inner product space.

For some time I thought that the less the math I study has to do with life, the better. The world of odd calculations and of shapes curved in impossible ways hooked me like a computer game. I visited Helsinki University for extra courses on topology and algebra—and of course on algebraic topology too! Oh the happiness of an intellectual feast!

I started to specialize in applied mathematics already when doing my engineer's thesis in 1993–1994. Although Olavi Nevanlinna, Jukka Tuomela and Olli-Pekka Piirilä sketched a quite theoretical topic for my thesis, I was happy to notice that it had connections to designing the movements of robots.

In 1994–1995 I did my alternative national service in the Rolf Nevanlinna Institute (RNI) of the University of Helsinki, which is a cradle of industrial mathematics. I got to know two men, who charted my scientific path: Erkki Somersalo, the great storyteller and a grand master of applied mathematics, and Matti Lassas, the most versatile thinker I know and since then my good friend.

My national service included teaching a special course on applied mathematics and mathematical models for mathematically trained people who were left unemployed by the big recession in Finland in the 1990s. It went great, because 14 out of the 20 participants found work after it.

An immortal conversation on the philosophy of science was left in my mind during my time at RNI. It was between the theorist Matti Lassas and the coding wizard Seppo Järvenpää. It went as follows:

Matti: How do you know what your program is calculating, unless you have proven your formulas?

Seppo: How can I trust your theoretical results, unless I've tested them on a computer?

Since then, my guiding principle in science has been: build computer programs that solve the user's problem efficiently and that are based on proven mathematical theorizing.

1.4 Thesis and Medical Imaging Research

For centuries, mathematics and the natural sciences were one. Only in the century of theoretical and conceptual mathematics, the twentieth century, they got badly separated from each other. I felt at once that the enthusiasm for physics in my youth and my technical experiments belonged together seamlessly.

I started writing my thesis with Erkki Somersalo as my supervisor at the Helsinki University of Technology in the January of 1996. The topic was not clear at the beginning, but there were many alternatives. In the end, a new breakthrough article by the Rochester University professor Adrian Nachman got chosen.

Nachman's work was on electrical impedance tomography. In this imaging technique, ECG electrodes are attached to the skin. If one wants to observe the functioning of the heart and the lungs, 32 electrodes are girded across the chest like a necklace of pearls. Harmless and painless electric currents are input through the electrodes according to well planned forms. Electricity flows into the body from some electrodes and out of it through others. The tomography machine measures the voltages on the electrodes.

Different human tissues conduct electricity differently. Electricity goes effortlessly through the heart as it is filled with blood, but electricity also readily bypasses the lungs that are full of insulating air. The currents seek their way from an electrode to another through the easiest path, and therefore inhaling and exhaling give totally different voltage readings. In principle, this makes it possible to make a picture of the divisions of conductivity in the body. The picture changes with heartbeats and breathing, and this gives the doctor a picture of the patient's internal organs.

Interpreting the electric currents into a picture is a hugely challenging mathematical task. The electric currents in the body don't move along known paths, but seek the easiest way through the organs. And it is the functioning of the organs that we want to know, so we can't use that knowledge in the calculation!

Nachman's article was the first to present a method for calculating an impedance tomographic image. But it's a terribly complex and technical paper. It's just the right kind of paper for over-eager young mathematicians to study.

I started reading in the spring of 1996. How bad could that article be? It's only 49 pages! Getting to know the introduction went quite well, as it gave an excellent overview of impedance tomography and earlier mathematical inventions on the topic. When I got into the first main section, my reading quickly stopped in a sentence on the top of page 15: "Here we are using the continuity of the Riesz transforms." What are these mysterious Riesz transforms? I used several months to find it out. Wading through the article and understanding it took me 2 years. Mathematical texts can be that dense and difficult to understand!

Nachman's work brought together a great amount of earlier research in surprising ways. Some of his techniques were developed for quantum physics to determine the colliding particles out of the splinters left by particle collisions. Some formulae were developed to explain the structures of tsunami waves, others to describe phenomena on the surface of charged particles, and a few to divide sound signals into pure oscillating wave forms. Only in mathematics can such different pieces of puzzle attach to each other so seamlessly and determine a way of imaging the internal organs of a patient with electrical currents.

I had steadfastly decided to realize Nachman's ideas in the form of hospital equipment. But the task was huge. Nachman had assumed that the measurements were made with an infinite number of microscopic electrodes that are perfectly accurate. I had to turn the theory into a computational computer program that is able to make images of the data collected through 32 electrodes, and the data necessarily contains errors that diminish its accuracy.

I got lucky. Marko Vauhkonen wrote a thesis on medical physics in Kuopio in 1997. I took a train there together with Erkki Somersalo and Marko's opponent Margaret Chaney. During the trip, Margaret told me about Jennifer Mueller, a fellow in her research group. Jennifer was interested in developing a new imaging method for impedance tomography. We were able to invite Jennifer to visit Finland with money from Erkki's project grant in the June of 1997.

Finland is not always an easy destination for international visitors. I took Jennifer to the HUT guest room at the Otaniemi campus. I showed her the way to the University main building, agreed to meet her at 9 o'clock in the morning and went home. At 10 o'clock I got a very angry phone call. "Where is everybody?" Jennifer asked. "I tried to go to the university, but the doors are locked." Poor Jennifer was confused by jet lag and had woken up after an hour's nap, thinking that the bright summer evening is the morning. She had been wondering about tech students having a barbeque, but had assumed that it had something to do with odd Finnish breakfast customs.

The visit had a bad start, but it led into a friendship and a scientific cooperation that are still ongoing. Over the years, we have developed a new imaging method for impedance tomography step by step. This has required modifying and expanding on Nachman's ideas, and realizing them through computing. One of the high points of our long project was a 2004 paper, where we showed that the model works for data collected at a lab. Another was a paper in 2006, where we reported that we had used the method to make medical images of a real person for the first time. Before that, we had only tested our methods on computational models. Oh the great feeling, when I saw a video of the heart and lungs of a living person, which was calculated using methods developed by our group!

We're still continuing the development work. Jennifer has built an apparatus for measuring people in her lab in Colorado. She concentrates on imaging cystic fibrosis, especially in children. They have mucus in their lungs that makes it difficult to breathe for them. Therefore it's important that doctors are able to estimate how much air gets in the lungs. One can use a spirometer to do the measurement with adults, but small kids don't know how to use one. Impedance tomography offers an easy way to find out, how well the lungs are.

I've brought together a Finnish research group to investigate the use of impedance tomography to image cerebrovascular diseases. There are two main types of diseases: a thrombosis or a blood clot in the brain prevents the flow of blood into a part of the brain, and a haemorrhage. Both have similar symptoms (a smile on the other side of the face, confused speech, an inability to raise the other hand on the top of the head), but they have completely different treatments. A thrombosis patient should usually be given a thrombolysis as soon as possible to dissolve the clot, but thrombolysis is dangerous for a person with a haemorrhage. Usually a haemorrhage can be found or ruled out with an X-ray tomography, but it takes precious time to get to the imaging device. With impedance tomography, one could find the cause of the stroke right in the ambulance.

1.5   As an Industrial Developer

After my doctoral thesis, I was hired by Instrumentarium Imaging to develop medical technology. I worked on three-dimensional X-ray imaging. I'll tell you about it later. Here I look back at joining the brisk Instrumentarium team in the corporate orienteering league.

Orienteering too is an adventure in the world of mathematics. A map is an image of the real world. Bodies of water, roads, buildings, rivers and other significant locations are drawn into it in miniature. The geometric accuracy of a map makes it a mathematical model: from my home it's 1 km to the station, and in the map (1: 10,000) my home and the station are drawn 10 cm away from each other.

My colleagues coaxed me into an orienteering competition despite my warnings. According to my wife, one should walk in the opposite direction to the one I'm proposing when visiting an unknown city. My friends didn't support me either, when I tried to deny such charges in the social court of my friends. I was no star player to aid the orienteering team. Except in the sense that a minimum headcount was required of the team, and without me it would not have been reached.

My first competitions went in learning the basics of orienteering maps. I for example learnt that a thick line with cogs and next to a lake is not a path, but a cliff rising right out of the water. Another thing surprised the bookworm from Töölö. The fastest route between two points is not always a straight line! Even the theorems of the great Euclid are twisted, if there is a spot marked by blue dotted lines between the points A and B. I went for a record time to reach the checkpoint and fell into a swamp up to my groin—my cell phone was of course in the pocket of my trousers.

After initial difficulties, I quickly reached the next level in reading maps. Of course, that had problems too. It helps to think of the path from one's position to the next checkpoint with contour lines on the map. First along the left side of a gentle hill, then there's a cliff on the right, and finally between two small hills onto low-lying land. The course is completely clear in the geometric ideal world of the map. On the ground, I saw only bushes and had no sightings of the rising and sloping shapes of the terrain.

In traditional paper maps, many things are correct, like the paths and the contours. Every twig and tree isn't marked on it, and it gives no X-ray vision for seeing through obstacles.

Digital maps like Google Maps are more advanced mathematically. You can use them to travel down to the streets of Barcelona and walk your way from your hotel to the Ramblas in the virtual world. You use photos to identify buildings and notice interesting shops and cafes. Google Maps is an unprecedentedly accurate model of the world, but it cannot tip you off about a garbage truck that has been parked in the middle of the street today and that blocks your favourite route.

Remember this rule: if you're lost, go back where you last knew your location. It's a good guide in research too.

2

Examples of Mathematical Models

Have You Ever Run Across the Pair of Words *mathematical Modelling*? Don't worry if it sounds strange. Most people haven't ever heard of them. That's how it is, even though mathematical models run our everyday life and the technological foundations of our society. Even the AI that will soon drive our cars and write our news articles is based on computational models of nerve cells.

Even if you happen to belong to the majority of people who find mathematical modelling strange, that won't be for long. The main goal of this book is to introduce and explain mathematical models and describe, how we professionals do modelling.

Mathematical models are a bit like making a sketch of Rembrandt's famous painting *Night Watch* with a ballpoint pen. The subtle changes of shades in the master's oil painting are of course left out, and the stunning dialogue of lights and shadows can't be duplicated with a blunt blue line. But you can count the number of people, note the drum on the right and see that the watchman on the left is loading his rifle even from a coarsely and shoddily dashed copy. A ballpoint drawing is an approximate model of the

© The Author(s), under exclusive license to Springer Nature
Switzerland AG 2021
S. Siltanen, *Step into the World of Mathematics*,
https://doi.org/10.1007/978-3-030-73343-8_2

original painting. The model catches some features of the original, and misses out a lot. This is how mathematical modelling works too. Useful models succeed in catching those features that are important to the user of the model. The model can be completely off in less important features.

Mathematical modelling is a difficult art, because it combines three very different things. First is that phenomenon in the real world, which one is trying to imitate but which limited human beings can only describe with approximate accuracy. The other is skill in computer programming, which resembles craft work. The third material for modelling is mathematics, that rigorous but powerful bastion of exactness.

In this chapter, I'll tell about mathematical modelling by using examples. I'll start with the example of human gender models, which are not mathematical but bring up the basic idea of modelling and some problems with models. The three next examples concern trade, a computer football game and the lighting of imagined spaces in computer animations. They are suitable for mathematical modelling, because the circulation of money, the flight of a ball and the movement of light rays all obey computational laws. Last I offer a bit more technical account of how one can build a mathematical model for the movement of a school of fish and program it on a computer.

2.1   How to Model Gender?

Dividing human beings into men and women lies deep in many of the cultures and languages of the world. Most European languages choose words out of two alternatives according to gender. Japanese has both a masculine and a feminine way of speaking, which differ in other ways than just their words.

This binary model of gender is based on biology through the sexual reproduction of our human species. According to it, the bearers of the chromosomes XY are men, and the combination XX means a woman. The division man/woman is used in passports, toilets and sports competitions. The model is simple and fits well into the tendency of our brains to see oppositions everywhere, like hot/cold, good/evil or cheap/expensive.

However, the dichotomy is a coarse gender model. It cannot describe even the plurality of biological sexes.

The division into XY men and XX women is not exhaustive. For example, some persons have the chromosomes XYY, XXY or XXX. One can try to force this into the dichotomous model by defining that everyone with a Y chromosome is a man and the rest are women. But even that won't help, because medicine knows of cases in which a part of a person's cells have the chromosomes XY and a part have the combination XX. What is their sex (Ainsworth, 2015)?

In addition to biological sex, it's good to take into account a person's own experience of his/her gender. Research has shown that over 90 out of 100 persons are happy with the gender they were born with. But about 5–10% of us does not feel that the gender on their passport is really theirs. This group of people includes transsexuals, intersexuals and non-binary persons.

We can build a more complex model by introducing a third possible gender: woman, man and other. This trichotomous solution is more complex but also more comprehensive than the model with two alternatives.

On the other hand, "other" is quite a simplification, because it means "everyone who does not represent the first two choices". Could we make the model even more accurate?

There are many phases in the development of a foetus that often go according to the binary man/woman model. But about a percent of born children are intersexuals, who cannot be classified as boys or girls according to the structure of their sexual organs. The possible outcomes of the development can be roughly divided into five alternatives (Fausto-Sterling, 1993).

If we take into account the chromosomes, the structure of the sexual organs and a person's own gender experience, we will need over a 100 genders in our model. Such an exact sorting will cause practical problems, if we want every gender to have their own toilets and categories in their passports. This is not a very good model any more, because it doesn't structure and simplify the modelled phenomenon enough.

The examination of models for human gender shows, how mathematical modelling usually proceeds. We start with a simple and clear model that explains an important part of the phenomenon. If the model is too coarse, we add features into it and accept that the model will be more complex. We keep adding details until the model serves its purpose. If the complexity goes too far, we return to a model that's a step simpler.

2.2   Natural Numbers Model Quantities

I know a person called Joonas, who recently made a strong claim: "1 + 1 isn't necessarily 2. That's a matter of convention." This interesting new idea teased my brain, but it also made me think about it more carefully.

Joonas is right, in a way: 1 + 1 can be something else than 2 if one agrees so. For example, when one calculates logical truth values, one sometimes uses 1 for "true" and 0 for "false". "+" too means the logical operation "and", and this leads to the formula "1 + 1 = 1". In other words, "true and true = true".

We also can take a look at integers "modulo 2", where only the remainder of dividing by 2 is taken into account. This funny world of calculations turns every odd number into a one and every even number similarly into a zero. Then 1 + 1 = 0, because two and zero are the same thing, "modulo 2".

I don't know if Joonas was talking about such odd but mathematically precise examples. If he was, he was in the central regions of pure mathematics. In these regions, one can use the imagination to construct any way of calculation that works consistently. It's a fantastic and profound valley of human culture, where ornaments have been developed over centuries and they compete with each other for cool gracefulness.

I suspect that Joonas meant the addition of natural numbers to be a matter of agreement. But it's not!

Addition is a mathematical model for the number and quantity of things. It is a part of applied mathematics, where the purpose of formulas is to faithfully and accurately mirror phenomena in the real world.

Imagine that you are standing in a queue to the Rocket in the Linnanmäki Amusement Park in Helsinki. If it's your turn and there is only one place left, you cannot solve the problem by agreeing that me + my friend = one person. There is no room for two, and you'll have to wait until the next ride.

A second example: you need a clothesline between two trees when staying at the summer cottage. The distance between the trees is 7 m. Suitable string is sold in rolls of 3 m each at the store for "thousands of goods", which is located at the parish centre. You can of course agree with yourself that 3 + 3 = 10 m and save money by buying only two rolls. Still there is only 6 m of string, and it's not enough for the 7 m between the trees. Not to speak of rolling the ends around the trees and making knots.

The third example is nuclear fusion. Many kinds of reactions produce energy at the centre of the sun. In one of them, a heavy hydrogen or *deuterium* nucleus of one proton and one neutron fuses with an ordinary hydrogen nucleus, or a proton. The result is a neutron and $1 + 1 = 2$ protons. The produced element lies in square 2 of the periodic table, so it's a helium nucleus. Even if we humans agreed amongst ourselves that $1 + 1 = 3$, the new element would be helium and not lithium, which lies in the third square of the periodic table. Then the agreement $1 + 1 = 2$ is more advantageous for understanding natural phenomena than the agreement $1 + 1 = 3$.

The goings-on in nature do not give weight to the wiggles of poor humans. An apple falls from the tree and waves hit the beach regardless of what we *Homo sapiens* think. We can try to understand natural phenomena to the best of our abilities and develop models for them by using the powerful but strict language of mathematics. We can make agreements and conventions, but they only hold between people.

Joonas, for this reason $1 + 1 = 2$. Not because we have so agreed, but because addition defined that way gives a strong, expressive and precise model for the numbers and quantities of things.

But even addition of natural numbers is an intermediate model, which has its defects. If there are both whole and half apples in the basket, we need *fractions* for successful additions.

It's possible to calculate a lot of cool stuff with fractions, but they too offer only an intermediate model. One cannot use fractions to solve the problem of calculating the diagonal (the straight line between the opposite corners) of a square with 1 m long sides. One needs the square root and the set of *real numbers* that comes with it.

Real numbers prove to be quite an intermediary model, for two reasons. First, one needs *complex numbers* to solve totally ordinary equations, and they bring the strange *imaginary unit* with them. Second, real numbers give us thought about a situation, where one can divide a billiard ball into five different parts and then reassemble them into *two* billiard balls, both with the size of the original! Though these five parts are shaped so strangely that their volume isn't even defined. This wonderful example by Stefan Banach and Alfred Tarski is so complex that I won't study it more closely.

2.3 A Model for Prices in an Online Store

Imagine that you're buying tires for your bike that cost 20 euros each. A quick addition tells that the total price for two tires will be 20 + 20 = 40 euros. If you're buying tires for a family of four, you can calculate 20 + 20 + 20 + 20 + 20 + 20 + 20 + 20 = 160 euros, or multiplying slyly 4 × 2 × 20 = 8 × 20 = 160.

Based on these calculations, it seems reasonable to draw a model for the prices of tire purchases:

Model for tire prices 1: total price = number of tires × 20 euros.

That works well for ordinary trips to the store. But what if you need more tires?

When building a model, it's good to think hard about the circumstances, under which the model makes sense. For example, if you plan to buy new tires to 1,000,000,000,00 0,000,000,000,000,000,000,000,000,000,000,000,00 0,000,000,000,000,000,000,000,000,000,000,000,00 0,000,000 bikes, *model for tire prices 1* acts as a faithful servant and reports that you should put 4×10^{91} euros on the table. But this doesn't make any sense: there isn't even an atom in the universe for each of the 10^{91} tires! And even Jeff Bezos, Queen Elizabeth and Jack Ma together don't have that much money! That *model for tire prices 1* is then quite an intermediate model.

Well, maybe that example of 10^{90} tires is a bit far-fetched. We mathematicians are often guilty of thinking about exaggerated examples. We can however draw a conclusion from it: we should set an upper limit to the number of purchased tires to the price model.

The model for prices also needs some extra fixes to take bulk discounts into account. if you buy 100 tires, you get a discount and the price is lower than 2000 euros. Let's say a 10% discount, so the price will be 1800 euros. Orders for 1000–10,000 tires also get a 20% bulk discount, for example.

Mathematical models are of most use in the form of computer programs. That's why I'm writing our improved price model in the kind of form an online bike store could use in its sales system:

Model for tire prices 2, suitable for a computer program
Mark the amount of tires a customer orders with an x.
If x is 99 or less, the total price is x × 20 euros.
If x is 100 or more, but less than 999, the total price is x × 18 euros.
If x is 1000 or more, but at most 10,000, the total price is x × 16 euros.
If x is over 10,000, tell the customer that one can't order so many tires here.

Model for tire prices 2 is more complex than *Model for tire prices 1*, but it imitates real trades better.

A mathematical approach is well suited to describing monetary exchanges. Money is just a means of trading that has been developed by humans, and it really has been already defined in a mathematical language, that is numbers.

2.4 The Flight of a Football on PlayStation

Can you imitate other spheres of life with mathematics too, and not just the economy? Let's take a look at the football games on PlayStation. They're starting to look so good that you can't tell a TV broadcast of real football game from a video game at a glance. At least that has happened to me; maybe people who are deeper into football can easily tell them apart. How is such a confusion possible? There's no ball in the console, not to speak of little players.

The tumult of the game on TV is similar to an animation. They show 30 pictures a second, and moving objects are drawn in a slightly different place picture by picture. Human eye is so slow that the brain forms the illusion of movement. There's an important difference between a game and an animated movie. All of the frames of a movie can be drawn beforehand. The program for the computer football game will have to draw the pictures as the game goes on, taking into accord the choices and movements the player sends into the console through the gamepad.

For example, if there is a free kick in the game, the computer shows a small animation of a ball rising to the air after being hit by the foot of a footballer figure. Next the position of the flying football must be drawn on the screen so, that the flight path looks the same as on a real football pitch. The calculation of the position of the ball can be done beforehand for a suitable collection of strengths and directions of the kick. Isaac Newton already invented the mathematical formulae for this. If one doesn't take drag into account, one gets an intermediate model where a ball flies on the screen on a parabolic path.

Many other things have to be computed in football console games, not just the location of the ball. The player figures must run, kick, make saves and fall in convincing ways.

This requires imitating the human anatomy geometrically and imitating walking, running and other movements naturally. The player using the gamepad directs one of the figures, but the other figures too must behave reasonably or like a real football team. All of this must be calculated according to mathematical rules 30 times a second.

2.5   Bringing Light to a Virtual Room

Let's focus on the interior design of a virtual space. How to place lamps in a room that only exists in the room of a computer game? It would be nice to try out the quantity and quality of light in different placements of lamps. In the real world, we can simply place the lamps, sink into thinking about the lighting situation with the head down and add a cord of LEDs behind the bookshelf as a cool detail.

But there are no lamps in the virtual room that doesn't really exist. It's made out of small imaginary and monochrome triangles. The computer tracks the x, y and z coordinates of their vertices. How can one model the lighting?

There are many solutions to this problem in computer graphics. *Ray tracing* depends on light rays, and it proceeds backwards from the eye of the spectator to the objects of the virtual room along every light ray. It's possible to use ray tracing to produce real looking reflections and lens distortions to the visible part of the room, like distorting an area seen through a fishbowl. Shadows become so sharp in ray tracing that they may even look unnatural.

The other main method for virtual lighting is *radiosity*, which I explain more carefully below. One must not confuse it with Radio City. Creating a lighting to a room with the radiosity method is based on choosing a colour to each triangle like colouring the numbered areas a children's colouring book. The triangles on a wall close to the lamp are brighter than shaded ones on a floor under a table. Two things affect the luminosity of every triangle: (1) the light emitted by the triangle and (2) the light coming from other triangles and reflected by the triangle. Of these, (1) has to do with the triangles in the parts of the lamp providing the light; other triangles reflect the light.

How does one select the colours of a triangle then? In the name of clarity, I keep to the world of a black-and-white game, where one has to select the grayscale for each triangle: the extremes are pitch black and snow white, and something in between for most triangles.

To illustrate this, I'm looking around my living room and estimate the number of triangles needed to model it. First, let's turn the floor into triangles. Let's place points on a floor like a rain squall showers drops on water to the asphalt on the parking lot in summer. We need a couple of points for each area roughly the size of the palm of a hand, so that there are enough triangles to guarantee the evenness of the virtual lighting. Next we connect each point to its neighbours with a straight line.

Now the floor is thoroughly covered with triangles of different sizes, like after a more artistic tile fitter. A living room floor of 20 m² will have about 4000 triangles. When the other walls, the roof and the floor are taken into account, we need about 15,000 triangles. And what about the table? A few 100 triangles the size of half a hand will be enough, and I'll form the coffee cup on the table out of a 100 of little triangles, for example. There are now about 15,500 triangles. My virtual studio is already making me go out of breath, so let's leave the bookshelf, the sideboard, the tile stove, the stereos and the Cajon drum set out of the model.

We of course need a lamp, otherwise we could just colour every triangle white and celebrate the quickest programming project in the world. My five-pointed ceiling light looks deceptively like Paavo Tynell had designed it. It probably requires about a 1000 triangles, each the side of a fingertip, to look about right in the virtual room.

I have now put together 16,500 vertices of different-sized triangles in a 3-D virtual reality so that the whole looks just like my living room. There's a virtual coffee cup on the table and a virtual lamp on the ceiling. It does not yet look lighted: all the triangles are the same, pure white. It looks like I had cleaned up a sack of wheat flour that had broken on the floor by using a powder-type fire extinguisher as a leaf blower.

We should add 16,500 grayscales to the triangles so, that the lighting of the real living room looks the same as the lighting of the virtual room on computer screen.

The calculations for radiosity lighting have to be done with a *system of equations*. Take the example of two triangles: one from the floor and the other from the wall of the virtual room. The luminosity or the grayscale of the floor tile is affected partly by the luminosity of the triangle on the wall. One has to take into account the distance between the triangles as well, because luminosity decreases proportionately to the square of the distance. One must also take into account the solid angle, in which the triangles see each other. For example, two triangles on the floor don't affect the luminosity of each other directly, because the triangles only see each other's sides. The luminosity of a floor triangle gets a small additional term (the luminosity of the triangle on the wall) × (a spatial geometric figure about their relative positions). But the luminosity of the wall triangle is unknown and depends on the luminosity of the floor triangle, and vice versa.

So we have to write a system of 16,500 equations, where the unknowns are the grayscales of the triangles, the coefficients are geometric features and the known terms are the luminosities of the triangles that make up the lamp. It won't take long for the computer to crunch these numbers, once these geometric coefficients have first been fixed. That's quite a job, which the poor participants on my modelling courses can undoubtedly confirm!

The calculation of radiosities is only an intermediate model. It doesn't know how to produce reflections out of glittering surfaces or mirrors, and the light rays don't bend correctly when coming through a transparent pitcher that is on the virtual table. These phenomena are better described by a ray tracing model, which models the paths of individual light rays, with their bends and reflections. Although ray tracing smacks of an intermediate model too, because it produces unnaturally sharp shadows. The radiosity method does soft-edged shadows better.

Box: Systems of Equations

Let's start with the simplest case, a *pair* of two equations.

Imagine a situation in the summer. An unknown number of jumping spiders, small predators of the family *Salticidae*, are preying on the sunny wall of a log cabin. A few viviparous lizards are warming up there too. We happen to know that there are 48 legs and 44 eyes dwelling on the wall. *Question*: How many spiders and how many lizards are there on the wall?

Let's mark the number of spiders with an S and the number of lizards with an L. Let's write a system of equations for the number of eyes and legs by using the following information:

- The jumping spiders have eight legs and eight eyes.
- The viviparous lizards have four legs and two eyes.

We get two equations, or a system containing a pair of equations. The first one is for the number of legs: $8S + 4L = 48$. The other equation gives the number of eyes: $8S + 2L = 44$. We cannot solve either of these equations alone, because they contain two unknowns (S and L).

We solve the equations through the *elimination method*. Subtract the second equation from the first, and we get:

$$8S + 4L = 48$$
$$-8S - 2L = -44$$
$$2L = 4.$$

(*continued*)

Box   (continued)

Then we know that L = 2. Substitute this piece of information into the first equation, and we get 8S + 8 = 48. We can solve first 8S = 40 and then S = 5. Therefore there are 2 lizards and 5 spiders.

There can be more than two equations in a system of equations. We know that there are S spiders, L lizards and G grass snakes lying lazily on a rocky islet. There are 104 legs, 98 eyes and 5 tongues on the island. The following biological facts are available to us:

- Jumping spiders have eight legs, eight eyes and no tongue at all.
- Viviparous lizards have four legs, two eyes and one tongue.
- Grass snakes have two eyes and one tongue, but no legs.

We cook up three equations out of these ingredients. From the number of legs we get 8S + 4L = 104. The number of eyes is 2G + 2L + 8S = 98. The number of tongues gives this: L + G = 5.

You can solve this system of three equations as an exercise. The solution is: S = 11, L = 4, G = 1.

We mathematicians use *matrices* to solve large systems of equations. I'll tell you about them in the video "Wonders of mathematics: Matrices" that is on my YouTube channel *Samun tiedekanava*.

2.6   How to Build a Mathematical Model

I'll build a model of a virtual school of fish and describe the necessary steps in the work. My goal is to build a 10 s computer animation, where a school of fish is swimming in a shallow pool as realistically as possible. The end result is a video file with 600 digital pictures; 60 pictures are shown every second. The *frame rate* is 60 hertz.

Every frame should have the same number of fish, for example 200. A single fish should also move so little between subsequent frames that the impression of natural movement is formed in the brain of the viewer. And of course the fish should move as a school, and not as individual fish treading their own paths. The school behaviour is the challenge requiring the most work in our modelling task.

In fact, the goal is not to make a single video clip. I'll use a mathematical programming environment to write general instructions, which allow the user to freely select the number, the location and the direction of the fish at the beginning of the video, and the measurements of the rectangular pool. It's possible to modify the behaviour of the school in many ways by using the instructions. After making all these choices the user can run my short program and get a new video as a result.

I use *Matlab*, the favourite programming environment for applied mathematicians, for my programming. The name is an abbreviation from the words *Matrix Laboratory*. (By the way, *The Matrix* is my favourite movie, although that piece of information has nothing to do with this.) Any mathematical programming environment is suitable for this job, like for example Python, Octave, R, C++, or FORTRAN.

Before we immerse ourselves in the world of a mathematical school of fish, we must keep in mind an important basic fact. One should build mathematical models according to certain rules. Or, there are really two alternatives, A and B:

Alternative A
 – Be enthusiastic and try to write a program that is comprehensive enough for the exciting result can be immediately seen in all of its magnificence.

- You find out that running the program ends in an error report.
- Fix the errors with confused setups.
- See that the program works, but the result could be anything.
- Modify the situation by adding a shortcut to the code.
- Fix the mistakes caused by the shortcut.
- You notice that your models produces pictures that have nothing to do with the modelled phenomenon, but could succeed as cover art for 1970s style psychedelic rock bands.
- Go through the above in random order.
- In the end you have to accept the fact that the whole work has to be started from the beginning, using the alternative B.

Alternative B

1. Make a very simple version of the model that describes only a tiny detail of the modelled phenomenon.
2. Test the model many times to make sure that it works right. In the best case, you can calculate the hoped output both with the model and by hand and check that the results are the same.
3. Is the tested model expressive enough? If it is, you're done. If not, add the smallest possible amount of new material to the model and return to instruction 2.

Every reasonable person chooses the alternative B and builds the new model systematically in the shortest possible time. But the temptation to pick the alternative A is great, as we will see in the following section.

One can model the movement of a school of fish speeding through the sea with three rules for individual fish. Together, these rules give a surprisingly realistic model for the school:

School Rule 1 Try to swim in the same direction and with the same speed as the nearest neighbouring fish in the school.

School Rule 2 Try to keep the same distance to your nearest neighbours, and if you stray from the school, swim towards it.

School Rule 3 Don't collide with other fish (Reynolds, 1987).

I want to keep my example of modelling virtual fish simple. That's why I'm modelling a school of fish in a shallow pool, because then the fish can only move in a two-dimensional and limited space. A more complex model would allow the school to swim in a three-dimensional sea that is limited from below by the bottom, from the above by the surface, and from the edges by the shore.

I'll start programming the school of fish from the stage 1 of the reasonable alternative B. The simplest model requires a clip of code for drawing the locations of the fish into the picture for each of the 600 moments. I divide each second to 60 moments, because this corresponds to the frame rate of the final video. In my thoughts, the imaginary school of fish is swimming in a square pool with a width of 20 m. I want to draw a view from the above, so there is a blue square in the picture.

My First Fish Program
- Draw a blue square to picture the pool.
- Place a single fish at the centre of the pool.
- Set the swimming direction "right" and velocity "0.05 m between the frames".
- Choose time = 1.

(a) Draw the fish into its current place.

- Move the fish in its swimming direction the same distance as is its speed.
- If time >600, halt the program. Otherwise add 1 to time and move to (a).

The program text above is not computer code, but plain English written to people. I think that's more suitable for a popular science book like this.

As a result of this energetic first step, I see a dot moving slowly to the right on the screen. A dot? Yes, I concentrate on drawing the locations of fish in the picture in the entire project; I won't even draw any object that looks like a fish. I'll leave it to the Hollywood animators, if my draft film gets wind in its sails.

But this fish in the first program swims right through the wall at time 201. My program doesn't take the size or even the existence of the pool into account. The model must then be improved.

Right now we can test the correctness of the code. If the fish moves 5 cm between the pictures, it takes 20 moments for the fish to move forward 1 m. Then swimming half a pool or 10 m takes exactly 200 moments, and the fish is outside the pool at moment 201. The fish is then moving from the right place in the right direction with the speed we wanted, so this works fine in my simple starting model.

I'll now add new features to the model, but only a few. I'll add the edges of the pool to the code, and I'll recklessly add one more fish. I'll use the x and y coordinates like I was looking the pool from the above so that the edges of the pool have the same direction as the coordinate axes. I'll place the origin, or the centre of the coordinate system, at the centre of the pool:

My Second Program for a School of Fish
- Draw a blue square (=the pool).
- Place one fish in the centre of the pool (x = 0, y = 0) and another at x = 2 and y = 1.
- Set the direction of the centre fish as the x-axis and its velocity "0.025 m between frames".
- Set the x and y velocities of the second fish as "0.025 m between frames".
- Choose time = 1.

  (a) If a fish is even a little outside the pool, move it inside the pool to the edge and reverse its swimming direction to point back inside the pool.

- Draw the fish into their current place.
- Move the fish in their swimming direction the same distance as is their velocity.
- If time >600, halt the program. Otherwise add 1 to time and move to (a).

Now the two fish are moving at a constant velocity, until they hit the edge of the pool and turn back. Great! The program is however boring as a movie, and one cannot find any school behaviour yet. I'll continue adding complexity to the model.

I want to see more fish in the pool, and they must be aware of the presence of other fish. All of the rules of school behaviour should not be added at once, because the alternative B instructs to add the least amount of features to the model at a time. I'll choose to implement the school rule 1 first: "Try to swim in the same direction and with the same speed as the nearest neighbouring fish in the school."

The Third Program
- Draw a blue square (=the pool).

- Place 100 fish in random places near the centre of the pool.
- Set the directions of the fish "to the right, but randomly a bit up or down."
- Set time = 1.

  (a) If a fish is outside the pool, move it inside the pool to the edge and reverse its swimming direction to point back inside the pool.

- Draw the fish into their current place.
- If a fish has neighbours closer than 1 m, change the direction of the fish towards the centre of the group of the nearest fish.
- Move the fish in their swimming directions the same distance as is its velocity.
- If time >600, halt the program. Otherwise add 1 to time and move to (a).

The situation is fine, as the rule seems to work. Namely, the fish seek out tight clusters that swim in the same direction. But the clusters start going obsessively clockwise around the edges of the pool, like the polar bears that have been disturbed by their imprisonment in a zoo! This doesn't look good, so we need other school rules as well.

Here I confess that I in fact followed the alternative A enthusiastically and wrote all the school rules into the code at once, without testing them enough one by one. The result was chaotic except for all of the fish seeking out the diagonals of the pool. What was going on? After a hard debugging session I noticed that my code gave the wrong results in situations, where the immediate neighbourhood of a fish had only one other fish and not several others. I would have saved time if I had acted in accordance with the alternative B and tested every detail carefully. Don't do what I do, but what I recommend in the alternative B.

Let's Next Test School Rule 2 "Try to keep the same distance to your nearest neighbours, and if you stray from the school, swim towards it." This is the fourth program, so I'll leave out the initial preparations. They're the same as before.

− Set time = 1.

    (a) If a fish is outside the pool … (See the previous program)

− Draw the fish into their current place.
− If a fish has neighbours closer than 1 m, change the direction of the fish towards the centre of the group of the nearest fish.
− If a fish doesn't have neighbours closer than 1 m, change the direction of the fish towards the nearest school of fish.
− Move the fish in their swimming direction the same distance as is their velocity.
− If time >600, halt the program. Otherwise add 1 to time and move to (a).

Now this is funny! No matter what the initial settings are, the fish quickly organize into a spherical formation with a diameter of 2 m. Inside the sphere they bounce around like angry hornets, who cannot find the creature that blocked their nest despite feverishly trying to find it. The cohesive force of rule 2 is clear.

The School Rule 3 "Don't Collide with Other Fish." Still Has to be Implemented. This is the Fifth Program Place 100 fish in random places in the pool. Set the directions of the fish "to the centre of the pool" and speed "0.1 m between frames."

− Set time = 1.

    (a) If a fish is outside the pool …

- Draw the fish into their current place.
- If a fish has neighbours closer than 15 cm, change the direction of the fish away from the nearest fish.
- Move the fish in their swimming direction the same distance as is their velocity.
- If time >600, halt the program. Otherwise add 1 to time and move to (a).

Now see what's going on! A disorderly set of points across the pool zooms towards the centre. The underwater traffic jam of all times is about to happen! Are the fish going to get bumps? No, because the strict implementation of school rule 3 throws fish in different directions just before the collision, as if asteroids in space were swishing in each other's gravity fields. The modelling is starting to show its strength, and the movie looks quite fascinating. Of course there's no school behaviour visible here, but we get it by combining all of the three rules in the same program.

This Is How We Get the Sixth, Ready Program Code
- Draw a blue square (the pool).
- Place 100 fish in random places near the centre of the pool.
- Set the directions of the fish "to the right, but randomly a bit up or down."
- Set time = 1.

  (a) If a fish is even a little outside the pool, move it inside the pool to the edge and reverse its swimming direction to point back inside the pool.

- Draw the fish into their current place.
- Change the directions of the fish according to a weighing of the three rules:
- school rule 1 (20%), school rule 2 (30%), school rule 3 (50%)

- Move the fish in its swimming direction the same distance as is their velocity.
- If time >600, halt the program. Otherwise add 1 to time and move to (a).

Now I can see a set of points on my computer screen, swimming around as a united front in its square pool. The group is confused for a moment after hitting a wall, but soon regroups. The location of each fish moves slyly to and fro according to the movement of the others. There's a sparser group moving at a constant distance on the edges of the larger group, as if it was a vanguard or an escort.

One could still improve the fish model by adding a bit randomness to the directions and locations of the fish on each round. This would prevent too regular formations from springing up, and model the turbulent currents in the water.

The set of points I created with a computational model does indeed bring a school of fish into mind! Its communal-looking movements invites the question, whether there's an officer in a command centre in the background, giving orders to each soldier in his fish army with a radio phone. But there isn't any, and each "fish" obeys three simple rules, which only require information about the places, speeds and directions of the closest neighbours.

That's how the model gives us information about the real world. When we see a flock of birds moving on the sky above Berlin as if it were an organism with a will of its own, we can easily ask, how does the leader of the flock communicate with its members and direct them to fly the right way. The model tells us that there is not necessarily need for such a central command. The model is no proof. It does not tell us that there is no central command. But it reveals that it is possible to produce behaviour that corresponds to real schools of fish by having each member follow locally three easy and natural rules.

And the model does not have to be a source of scientific knowledge, even though it can serve as such. For example, for an animated film it is enough to have a result that just looks good.

You can learn more about the movement of a computational school of fish in the video "Mathematical modelling: School of fish" on *Samun tiedekanava* in YouTube. The codes are available at https://github.com/samuntiede/school_of_fish/.

References

Ainsworth, C. (2015). Sex redefined. *Nature, 518*(7539), 288–291.

Fausto-Sterling, A. (1993). The five sexes: Why male and female are not enough. *Sciences-New York, 33*, 20–25.

Reynolds, C. W. (1987). Flocks, herds and schools: A distributed behavioral model. *Proceedings of the 14th Annual Conference on Computer Graphics and Interactive Techniques—SIGGRAPH '87. Computer Graphics, 21*, 25–34.

3

The Big Models of Earth and Space

Science Is a Fruit of man's unquenchable thirst for knowledge. The desire to understand has created the entire scientific culture, the universities and research centres and innumerable ways for finding out, how things are. Mathematical modelling or *computational methods in science* is one of these methods. In the last decades, it has risen as a new way of seeking the truth alongside theory-building and empirical research.

The most interesting questions cannot often be clarified with direct observations. Will it rain next Sunday? Or next Midsummer? The average temperature of the Earth has been rising rapidly for a long time: how hot will it be here in a 100 years? What happens when black holes collide in space?

Mathematical models offer answers to these questions. But they're just intermediate models, so the answers will always be approximate too. In this chapter, I'll tell which ones of these questions we can investigate with models and which ones we can't. Why climate change can be predicted far into the future, for example, but the weather can't?

© The Author(s), under exclusive license to Springer Nature Switzerland AG 2021
S. Siltanen, *Step into the World of Mathematics*,
https://doi.org/10.1007/978-3-030-73343-8_3

3.1 Weather Forecasts, Atmospheric Models and the Butterfly Effect

Dividing the World Into Boxes

Weather forecasts are based on a mathematical model of the atmosphere. The atmosphere and the oceans are divided into boxes in computer memory. The boxes are adjacent to each other like sugar cubes in a package. In each box, a book is kept of temperature, wind direction and strength, atmospheric pressure, water vapour, the amount and nature of atmospheric aerosol particles, clouds, solar radiation and other observables. The boxes in the sea have a book of currents, temperature, salinity and so on.

It is as if one had put an imaginary weather station in each box. The station sends a weather report to the headquarters at regular intervals, e.g. once in a minute. The computer CPU plays the role of a headquarters in the atmospheric model.

The model drafts a forecast step by step. First data about temperature, pressure and other data at the time of measurement is input into every box. Then the model calculates the physical phenomena within the box and the interactions between the boxes that take place during 1 min. When the forecast for the next minute has been drafted, the model takes it as the new "current situation" and calculates the forecast for the next minute. The result then is the weather situation 2 min after the initial time of the calculation. This can in principle go on forever, going 1 min to the future at a time.

When one is making a two-day weather forecast for Europe, one has to pile computational boxes only over our own continent. 2 × 2 km is usually chosen for the bottom and 1 km for the height of the box, for example. The idea is that the situation within the box is completely homogenous, or the temperature, pressure, humidity and so on are completely the same in the whole box.

Why take this coarse 2 km × 2 km × 1 km modelling and not boxes with a side of 10 m, for example? It would be great if the model had a few boxes for my home yard and one for the neighbour too. The reason is that the computer program would become too heavy.

The area of Europe is 10,180,000 km². How many boxes with a 2 × 2 km bottom do we need to cover the continent as if we were covering it with big pavement tiles? The area of the bottom is four square kilometres, so the answer is 10,180,000 divided by four. That's 2,545,000 slabs. They're piled up in 30 layers, and that takes over 76 million boxes. Moreover, the atmospheric model needs 2880 1-min steps to get 2 days into the future. The computer must then crank out as many as *220 milliard* (in the US, billion) imaginary weather reports.

That 220 milliard boxes is an incomprehensibly big number, and processing the reports it is difficult even for a modern supercomputer. But if we moved to 10 m × 10 m × 10 m boxes, the size of the model would be multiplied even more. There are 4 million little cubes in a 2 km × 2 km × 1 km box, so we would have *a trillion* computational units instead of 220 milliard ones. A trillion has 18 zeroes and so much to chew that one should not input it into a weather forecast computer.

If one wants to forecast the weather for a longer period of time than 2 days, one has to use a global model. The whole earth has to be covered with computational boxes. The best of these models, the colourfully named ECMWF, the boxes are 10 km × 10 km × 1 km. One has to pick bigger boxes, because covering the whole earth with 2 km × 2 km boxes would jam the computer. The ECMWF model calculates a two-week forecast twice a day.

Determining the Initial Conditions

The atmospheric model is fixed onto reality with measurements: some boxes have a real measuring station alongside an imaginary one. There are a few 1000 professional weather stations across the world, and their quality varies. The best is the SMEAR II station that was developed at the University of Helsinki. It measures 1420 different observables: for example, concentration of greenhouse gases, the amount and quality of atmospheric aerosol particles, plant photosynthesis, and soil temperature, humidity and its nutrient content. Professors Pertti Hari and Markku Kulmala started developing the SMEAR stations already in 1989. Because each SMEAR II-station costs 10 million euros, there is only one in Hyytiälä and three stripped-down models placed in Estonia, Russia and China.

A network of hobbyists' research stations gives data from over 10,000 weather stations across the world The data is not of such high quality but covers more areas. One gets data from the seas too: from buoys and along sea routes. This data is very sparse; there are a few buoys here and there, and ships sailing across the seas report the weather at different times and places. Weather balloons are sent to the air across the world at regular intervals, and their measurements end up aiding weather models. Weather satellites take pictures of the earth all the time, and they help us to calculate the data that is needed in the weather models.

So we can get up-to-date measurement data for the initial moment of our weather model for many computational boxes, but not to nearly everyone.

The Flow Model and the Physics Model

Once the initial state has been determined with measurements, at least approximately, it's time to let the model make predictions.

This is done with two different processes of computation. The *flow model* describes how the situation changes from one moment to the next. The *physics model* on the other hand takes into account the effect of the modelled observables on each other during each computational minute.

The flow models are based on conservation laws. For example, matter or energy should not come into being from nothing or disappear, and momentum should be conserved. The air pressure on the surface of the ground or on sea should not change in strange ways. The temperature and humidity of the air should be kept in accord to air currents.

The flow model contains many common-sense features. For example, if a 100 l of air is blown out of the eastern wall of a box each minute, the neighbouring box must get the same amount of air through its western edge during the same period of time. Events between boxes are described by writing equations like Newton's second law, or force = mass × acceleration; the ideal gas law that describes the compressibility of air; and also friction with the ground and viscosity in suitable forms. Viscosity describes a gas' or a liquid's ability to resist flow. For example, syrup has a higher viscosity than water.

If you have run across the concept of a *derivative* at school, you can make the following mental note. The flow model describes the change of the atmosphere over time, so the flow equations include *time derivatives.* This is because derivatives are a computational tool developed to handle rates of change.

A phenomenon called *turbulence* makes modelling flow more complicated. One can observe turbulence at home at a water tap. If you let a narrow but continuous stream of water into a pot, the situation is very calm. But if you turn the tap to the maximum, a bloody confusion of water starts in the pot! The masses of water rage around in roaming little balls of differing sizes, and one can't predict the height of the water at a given point. The water rises and falls completely randomly.

Modelling flows has ended up on the US Clay Institute's list of mathematics puzzles, because it's so difficult. The list includes seven exceptionally difficult problems that mathematicians haven't been able to solve during decades or even centuries. The Institute gives a million-dollar prize for solving one of these problems to lure mathematicians into solving them.

One of these seven challenges is to give a proof that the Navier-Stokes equations for flows have a unique and well-behaved solution. Or alternatively give a counterexample that the equations have either many alternative or an especially ill-behaved solution. This "ill behaviour" could for example be a spoon stir starting an eternally raging chaotic movement that forms ever smaller mini-hurricanes in a coffee cup.

The complexities of turbulence and the gaps in our mathematical understanding of it make the atmospheric models into intermediate models that have their inaccuracies. Nature offers us turbulence in the real world, even though our models couldn't completely describe it.

The *physics model* has to describe phenomena connecting the physical observables of different boxes of the model at each computational moment. These include cloud formation, the effect of ice on the reflection of light, energy coming from the sun and the behaviour of radiation when it hits the atmospheric aerosol particles.

In the boxes of the *atmospheric models* the modelled clouds and aerosol particles *scatter* a part of the rays of the sun, that is they change the direction of the rays. A part of the radiation returns to space. Then the boxes below receive a little less light and heat than in a situation with no clouds and no aerosol particles. One can note this on a sunny day, when one ends up in a shadow of a cloud and a T-shirt isn't enough anymore.

Clouds are made out of small drops of water and ice crystals that are born when the air receives more water vapour than it can contain. They're extremely important when modelling the climate, because they reflect solar radiation on the one hand, and on the other hand prevent thermal radiation on the surface of the Earth from escaping into space. The physics model should predict the lifecycle of the clouds from their birth to rain and also their shape. The predictions tell, if light can pass between clouds or if they form a grey slab covering the entire sky, like in Finland during November. The job is more difficult, because clouds are smaller than the computational boxes even though the model assumes every box to be completely homogenous.

The Butterfly Effect

Chaoticity or the phenomenon known as the *butterfly effect* makes predicting weather even more difficult. It is sometimes turned into the following form: "There was a tornado in Närpes, because a butterfly flapped its wings in Patagonia." This isn't an exactly fruitful way of looking at it, the following is a better one:

> The chaoticity of the weather means that even small differences in initial conditions lead to huge differences after some time. Let's make a thought experiment. We have two Earths. Their weather is exactly alike on 1.1.2019 except for that butterfly in Patagonia. Both Earths have the same atmospheric properties like wind, temperature, air pressure and humidity. But one Earth has a butterfly flapping its wings in Patagonia, while on the other the butterfly stays still.

During the first few hours we can see no differences between the two Earths, and maybe not for a couple of days. But after a couple of months, at most, we see that one Earth has a completely different weather than the other. It rains in Seattle 1 and the sun is shining in Seattle 2, Tokyo 1 has a thunderstorm and Tokyo 2 has sparse high-level clouds, and a tornado rages in Närpes only in one of the cases.

The repetition of differences of initial values leads to the fact that one can reliably predict the weather only for about 10 days. Predictions for a week aren't completely correct either. There aren't enough weather stations and satellite pictures feeding measurements to the weather models to cover every computational box. Many butterflies could be flapping their wings even though no measurement could measure the resulting air flow.

It goes approximately like this: the more square metres have a weather station and the more accurately they measure, the further into the future we can predict the weather. The current network of weather stations and the accuracy of atmospheric models are enough for accurate predictions for the next 10 days, at most. If we had 10 times more weather stations and somehow got to use more powerful supercomputers, we could lengthen the period the weather forecasts can cover accurately. But we can't get accurate reports for an infinite long time because of chaoticity. Even if we could fit a weather station into every cubic millimetre of the atmosphere and could fit ourselves in somewhere too, the measurements necessarily have a *finite accuracy*, that is small mistakes. These mistakes amplify as the model goes on for longer periods of time, and it could again miss a tornado.

You can find out more about the butterfly effect by watching the video "Wonders of mathematics: the butterfly effect, weather forecasts and chaos" on my YouTube channel *Samun tiedekanava*.

3.2   Modelling Climate Change

The average temperature of the earth has been recorded for over a century. At the beginning there were fewer measuring points, but today there are 1000 across the planet. The climate has warmed rapidly since 1970, and the last decade has been the hottest in measured history.

Human induced global warming is a serious problem, possibly a question involving the destiny of the human race. How could we know how the temperature curve continues to develop in the future? An especially important question is, how can we assess the effect of alternative courses of action on climate?

From a Weather Model to a Climate Model

If we want to predict the state of the climate after a 100 years, for example, we have to include enough windows of half an hour so that they add up to a century. We need 1,752,000 periods of half an hour. There are so many virtual weather reports for the model to calculate that we need bigger boxes than the 10 km × 10 km × 1 km of the weather model. Usually the size of the boxes is 100 km × 100 km × 1 km.

The flow model for climate modelling works just like it does in models for weather forecasts.

The physics part of the climate models have many of the same features as the weather models: they mathematically describe cloud formation and the effect of ice on the reflection of radiation. One has however to include longer-term phenomena, which don't have an effect on weather forecasts for a couple of days, but which are essential in scenarios that take decades. These phenomena include the carbon cycle of plant photosynthesis, the effects of other hustlings and bustlings of organisms and methane that is released from swamps. It's difficult to predict the effects of melting ice caps on sea levels and the strength of sea currents.

The amount of energy coming from the sun depends on the amount of water vapour and aerosol particles in complex ways. Hanna Vehkamäki is a physics professor, who knows radiation and aerosol particles thoroughly. She told that once she was leaving on a holiday trip, but a question from her kids endangered her holiday plans: "Why does the other side of the river look gray and blurred?" Hanna answered: "Do you really want to know? The answer is based on aerosol physics and takes many hours to explain." Although the children were exceptionally enthusiastic about science and would have loved to listen, Hanna's own desire was to let go from work when she was on a holiday.

One of the big factors in climate models is the release of methane from swamps. Does it escape by bubbling? What's the impact of the metabolism of plant roots on it? Even a small difference of metabolism efficiency is multiplied by the number of plants! So biology too is important in improving the atmospheric models.

The Sources of Errors in Models

There are about 50 different atmospheric models across the world, and each of them is only an intermediate model with its inaccuracies. Both flow and physics models contain two types of sources of errors.

The first of them is approximations that are caused by too big computational boxes. Smaller boxes could improve the accuracy in principle infinitely, but in practice the memory of computers and the patience of researchers has its limits. For example, if we decrease the size of the boxes from 100 km × 100 km × 1 km to 50 km × 50 km × 0.5 km, every old box is divided into eight smaller boxes. Then there are eight times as many virtual weather stations in the model, that is over 3 million stations. Running the model on a computer takes *ten times longer* on a model with the accuracy of 50 km than it does on a 100 km one. And if we decrease the blocks even more, no computer can handle the necessary amount of data.

The second source of error is our own ignorance. We don't have a thorough knowledge of all the physical and chemical phenomena that are in the models. The sparse grid of measurement stations leads into ignorance about all the necessary concentrations of gases and distributions of aerosol particles. These problems can be solved by doing more research and investing in new SMEAR II measurement stations.

The treatment of clouds is problematic in the models too. It should be made to work in models with boxes 100 km wide, even though the clouds are much smaller than the boxes. The inaccuracy in cloudiness and its changes is the biggest uncertainty in climate change predictions.

In any case: these kinds of deficiencies in climate models always cause uncertainties in predicting the strength and speed of climate change. Therefore a honest researcher cannot answer the question "How much will the average temperature of the earth change during the next century?" with a certain and precise number.

One can use inaccurate atmospheric models to estimate global warming by *calculating the uncertainty into the answer.*

The inaccuracy of a climate model can be estimated by calculating the prediction many times and feeding different numbers for the poorly known observables on each run. Then we get a collection of predictions that represent different extreme scenarios. For example, we can then calculate averages out of them, and averages are a lot more reliable to measure. We can also calculate the standard deviation to go with the average. The standard deviation gives a statistical estimate, how much the predicted temperature can deviate from the average.

The Force of Averages in Predicting Climate Change

Different climate models make different predictions over the next 100 years. It's possible to calculate an average of the predictions given by all the models. As we can see from the dice example in the infobox below, the average varies a lot less than the result of single die rolls, that is climate models in this case.

But is the average reliable, even if all the models are wrong? Yes, it is.

Joseph Fourier's pioneer work increased our understanding of the thermal balance of balance already in the nineteenth century, and Ludwig Boltzmann and Gustaf Kirchhoff built upon it. John Tyndall showed that carbon dioxide captures thermal radiation, and Svante Arrhenius put all the results together in the late nineteenth century. The research went on in the twentieth century, and Syukoro Manabe's research in the 1960s and 1970s at the latest made the carbon dioxide and water vapour models very accurate. After that, our knowledge has only grown even more.

All of this physical knowledge has been verified countless times and written into the models. Accurate and experimentally verified mathematical descriptions of other phenomena have too been written into them as far as computational capacities allow.

The atmospheric models for weather forecasts use 2 km × 2 km × 1 km boxes and one-minute steps. In long term climate models the box is 100 km × 100 km × 2 km and the step is 30 min. In this sense, the weather forecast models are much more accurate than the climate change models both in terms of space and of time. Why are we able to predict climate change during the next 100 years, when weather forecasts for the next 2 weeks are inaccurate?

We can see the force of averages here, and I use an example with dice to illustrate it.

Predicting a single throw is just a matter of luck. Despite this, we can predict that the average of a 1000 dice rolls will with a very high probability be between 3.4 and 3.6. This accuracy of prediction is due to a mathematical truth called the *law of large numbers*, which tells that if a random phenomenon like a roll of dice is repeated many times, it agrees with a reliable statistical rule.

Box: The Force of Averages

Here are the results for my 10 throws of dice in a row:

5	3	6	2	1	1	2	6	2	2

The numbers swing randomly to and fro. You can see the same numbers as a graph in the uppermost picture of the graphs on p. (68?).

This is how I fill in the next row of numbers: I roll five dice at a time and get the numbers 6, 1, 1, 5, 1. I calculate their average: $(6 + 1 + 1 + 5 + 1)/5 = 14/5 = 2.8$, and I mark this number to the first column.

In the second column I get $(6 + 4 + 3 + 3 + 1)/5 = 17/5 = 3.4$, and I fill the rest of the columns the same way.

2.8	3.4	2.2	3.6	2.6	3.6	3.4	3.8	4.2	4.2

The extremes of 1 and 6 are missing from the row of averages. Of course it's possible in principle that the result of five rolls would be 6, 6, 6, 6, 6, and the average would be $(6 + 6 + 6 + 6 + 6)/5 = 6$. That just didn't happen above. It would be quite rare, as everyone who has played Yatzy knows.

For the third row I set the computer to randomly roll a 1000 dice and calculate their average. I repeat this 10 times and get the following row of averages:

3607	3592	3539	3540	3434	3553	3515	3543	3433	3562

This row is even more uniform. No number is less than 3 or over 4.

(*continued*)

Box   (continued)

This is due to the mathematical *law of large numbers*, which is an expression of the force of averages. It reads like this: the more dice are rolled, the more probable it'll be that the average of the results is close to 3.5, which is the average of 1, 2, 3, 4, 5, 6: (1 + 2 + 3 + 4 + 5 + 6)/5 = 3,5.

What kind of game of dice is the weather then? There's a party every year on the 1st of May on Ullanlinna Hill in Helsinki. On some years the sun shines and the temperature is 19 degrees Centigrade, and on others it's raining sleet and the temperature is zero degrees. But the average temperature in Helsinki in May has been between 5 and 15 degrees for the last 118 years. So taking an average over 31 days stabilizes the measurements of temperature. The average temperature of the earth is similarly a lot more stable observable than the weather on a given day of the year.

So the models represent the best available knowledge we have, despite their uncertainties. Even when we take the possible ranges for the uncertainties into account, it's clear that our planet is heating badly. There are good reasons to fight global warming with all our strength.

3.3   How to Observe the Collisions of Black Holes?

Albert Einstein's *general theory of relativity* revolutionized our understanding of gravity. He described the universe as a four-dimensional rubbery lump, to which the stars, planets and ants too cause dips with their mass. Three dimensions are used to describe location, and time is the fourth dimension. Light rays and rocks floating in gravitational fields move along curved tracks in this uneven background material, just like a marble ball changes its direction when rolling past the floor sink in a bathroom.

The equations of the theory of relativity have been tested time after time. The bending of light due to gravity in a "gravitational lens" was observed in 1979, and GPS navigators would lead us astray without corrections that are based on the theory of relativity. We have already been able to observe black holes indirectly, when they rip nearby stars apart. They were even able to photograph the glowing ring of hot matter falling to a black hole in April 2019.

But if the Universe is made out of four-dimensional rubber, shouldn't a big star crash make it tremble? Einstein predicted these gravitational waves already in 1916, but one was able to observe them only in 2015. The delay is due to the fact that the space-time stretching caused by the gravity waves is so subtle and insignificant so that detecting it will take mankind's measurement technologies to the limit and even a bit beyond it. We need a proper dose of mathematical modelling in the mix too!

The phenomenon producing gravitational waves must be shockingly powerful, because we observe the waves here on Earth only after they've weakened during travelling vast distances. On the other hand the event can't be close to us either, because then we wouldn't survive to measure them in that case.

A wildly spinning neutron star is an example of a source of gravitational waves. They're only made out of neutrons. Ordinary matter is made out of atoms, whose tiny nuclei contain protons and neutrons. There's a shroud of electrons around the nucleus, and it takes a much bigger volume than the nucleus: the diameter of the nucleus is one hundred thousandth part of the diameter of the atom. If the atom were the size of the Finnish Parliament, the nucleus would be a blob that is barely the size of a millimetre. So the atom is mostly empty.

But a neutron star is made only of this nuclear material, and that's why its density is incomprehensible. A handful of a neutron star weighs the same amount as Mount Everest.

A completely round neutron star doesn't cause gravity waves. There must be a mountain range on top of it, making it asymmetric. Like everything else to do with neutron stars, the neutron mountain range is exceptional too. For example, a 1 mm wrinkle counts as a mountain range on a neutron star with 8 km diameter. It's big enough to buzz vibrations into the gravitational field when spinning around.

Another event that is massive enough to cause gravity waves is the collision of black holes. That's exactly what they managed to observe with a measuring device called LIGO.

The people who found gravity waves were awarded with a Nobel, and not for nothing. The project took human scientific knowledge to the limit, and mathematical modelling played a large part in it.

There are two four-kilometre tunnels at a straight angle in the LIGO device, and a laser beam runs in both of them. Mirrors reflect both beams many times so that each beam is reflected almost 300 times before it is measured.

In the ordinary case, when space-time is calm and gravity waves don't rock it, each beam completes its trip exactly at the same time. But when a gravity wave hits, one of the tunnels has a different length than the other for a moment. Then the two laser beams won't match completely. Light is made out of electromagnetic waves, and the small difference between the two beams is measured with an *interferometry* or measurements that reveal, when the peaks and valleys of light waves don't match.

But the twists and turns of the space-time of the Universe change the distance of the mirrors in LIGO devices very little: *only one ten thousandth of the width of a proton.* This is like measuring the 40,000,000,000,000-km trip from

Earth to Proxima Centauri to the accuracy of a hair's breadth. What kind of measurement could pick that up?

The interferometry in the LIGO device is based on using a problem in quantum computing. Namely, even a very small error will throw off the quantum computer from a state that's advantageous for calculations. This is the biggest obstacle in the development of quantum computers. The LIGO device turns the problem to its advantage. The sensitivity of the interferometer is based on the fact that a quantum computer of one *qubit* is disturbed by the failure of the beams to meet just enough that its "calculation state" is ruined. I put "calculation state" in quotation marks, because it's not the purpose of this quantum computer to calculate anything. It is used only to detect the disturbances.

The LIGO device was then miraculously used to collect the most sensitive data about the bending of the Universe. When put in sound, the scream of the Universe sounds like this: "Chirp." Mathematical modelling then tells, how this cosmic twitter should then be interpreted.

Special *wavelets* were developed especially for the task of measuring gravitational waves. They help us to cook the pure and noiseless song of space-time out of the data.

Box: Wavelet Transforms

A digital sound file can be divided into parts in a clever way by using a wavelet transform. The two biggest wave-forms imitate the long-term changes in the sound. They're called the father and mother wavelets. The transform also uses half-size wavelets that are the shape of the mother wavelet. Also a quarter-, an eighth-, a sixteenth- and even smaller-sized wavelets are part of the transform. It's a bit like writing musical notes.

Sounds of differing pitches and durations are put handily into their own boxes in the transform, so that they can be separately strengthened, weakened or even removed. The

(continued)

Box (continued)

smaller wave-forms aren't needed in the parts of the recording that contain no fast changes like cymbal chimes or sharp snaps. They can be left out without the sound quality suffering. Efficient wavelet compression is based on this fact.

One of the most important developers of wavelet transforms is Ingrid Daubechies, the daughter of a coal mining engineer in Belgium. She's one of the most cited mathematicians of all times.

When she was developing wavelets in the 1980s, Daubechies was forced to step out of the mathematicians' comfort zone of theoretical structures and infinities. A collection of independent wave-forms and even formulae for using them had already been formed in the ideal conditions of mathematics. These formulae were formulated in a form suitable for signal processing as *filters*, which are rules for combining the samples of a signal with neighbouring samples in the right percentage ratios.

But before Ingrid these filters were infinitely long! This doesn't bother the theoretical mathematician. But if a guitarist presses on the *overdrive* in a digital effect device and has to wait forever for the sound to break, even the most faithful audience might go home in the middle of a gig.

The wavelet family developed by Daubechies is a revolution in signal processing: it's made of wave-forms, whose calculations can be done quickly with short filters (Burke-Hubbard, 1998).

But about what event in the Universe did this wavelet-purified "chirp" tell us about? We wouldn't know about that either without advances in numerical mathematics.

Einstein's complicated equations describe the curved space-time of the Universe with an incredible accuracy, but they're difficult to solve. One can handle only oversimplified situations when calculating by hand. It was only in the 1990s when computer programs were developed to approximately solve Einstein's equations. This made it possible to create an odd library of sounds. The researchers modelled two black holes with a computer, one with 20 solar masses and the other with 18. The black holes were made to collide in the virtual reality of the model, the Einstein's equations were solved and the approximate shape of the gravitational waves on Earth was saved on a hard drive. The same test was run on black holes of 21 and 18 solar masses, on the pair of sizes 19 and 17, then on pairs of 15 and 31 masses, and then on others. A wave-form "fingerprint" for all kinds of collisions between black holes was then stored in this manner.

The "chirp" of the LIGO device was then measured by using an unstable quantum computer and a clarified wavelet transform. The resulting sound was compared with a sound library playing black hole collisions by using the compressed audio processing that was developed in the 2000s. This combination revealed that two black holes of about 30 solar masses each had collided and combined about 1.3 milliard light years away from Earth a long time ago. Science is even more fantastic than fantasy.

3.4 Mathematics and the Wicked Problems of Mankind

Science and technology have produced unprecedented wealth, peace and quality of life to mankind. Poverty is decreasing all the time, and there haven't been as few wars ever as they are now. The use of efficient technology has also brought unprecedented problems for us to face. Here are the two most wicked: nuclear weapons and climate change.

The foundation of science and technology is in mathematics. Therefore we need mathematicians both in building a better future and in stopping serious threats. Let's think a bit of the role of mathematics in understanding nuclear weapons and climate change.

Limiting the spread and use of nuclear weapons depends primarily on political means. But mathematics is needed for finding solid knowledge to support political decision-making. For example, nuclear tests can be detected and located by using the global network of measuring devices that are used to observe earthquakes. Mathematical computational inverse methods are also needed to infer, where is the origin of the tremors shaking the Earth's crust.

A second inverse problem is at the core of nuclear weapons control. In 2017, the International Atomic Energy Agency IAEA approved a new kind of device to monitor that used nuclear fuel is not stolen for weapons use. The device is based on the same idea as medical isotope imaging or positron emission tomography (PET). A doughnut-shaped camera is placed around uranium rods, and it measures radiation coming from different directions. It's possible to use mathematical computational inversion methods to make a picture of the bundled fuel rods, and the picture will clearly show if nuclear material has been stolen.

What about climate change? I told about it above, how mathematical atmospheric models can predict the development of the Earth's average temperature after different actions to reduce greenhouse gas emissions. But we can't plan these reductions without mathematics, modelling, calculating risks and understanding magnitudes either.

For example, our choices about how to produce energy are connected with climate change in complex ways, and mathematics helps in thinking about them.

The use of coal and fossil fuels produces over a third of global greenhouse gas emissions. In addition, over a million people across the world die each year because of the air pollution that comes from burning coal. This is why we have a reason to replace these polluting forms of energy with ones that pollute less.

Under current technology, wind and solar energy cannot replace coal on a large enough scale. Their electricity production changes too much depending on the weather, and we don't yet have a sufficiently efficient ways for storing enough power. Electricity has to be used when it's fresh.

Nuclear power has a lot of good properties for limiting climate change, like efficiency, steady rates of production and negligible greenhouse gas emissions.

Let's compare the efficiency of coal and nuclear power. Burning a carbon atom creates four electron volts of energy. Splitting a uranium nucleus on the other hand releases as much as 200 million electron volts. Then a nuclear plant releases over *50 million times more energy* out of each atom than a coal plant, and that's why one needs to produce and transporvt a lot less nuclear fuel than coal.

That 50 million is such an astonishing number that I'll illustrate it with an example of an electric car. Imagine that we burn enough carbon atoms to move the car for a millimetre. Then splitting as many uranium atoms would move the car for 50 km, like from Helsinki to Porvoo.

The structure of a nuclear plant is like a big water boiler. It heats water at a steady power, and therefore one doesn't need to store electricity.

The carbon dioxide emissions of a nuclear plant are formed during building the plant, mining the uranium and transporting the fuel. Producing nuclear power doesn't create any harmful emissions at all. Taken together, the carbon emissions of nuclear energy per a produced unit of energy are as big as in wind power and less than in solar power (Bruckner et al., 2014).

What about the drawbacks of nuclear power? They include at least nuclear waste and nuclear accidents. Used nuclear fuel is toxic for over a 1000 years. Its radiation however decreases over time, and if storing the nuclear waste in deep underground repositories succeeds as planned, in the end the waste has cooled down quietly and become harmless. But how great is the risk that the waste starts to spread due to natural forces or human actions during the cooling process? And if it does, how bad will the consequences be?

Nuclear accidents are frightening, no doubt because of the invisible danger caused by the radiation. On the other hand they are rare, and even the worst ones haven't caused a lot of deaths. From a statistical point of view, wind and solar power in fact cause more deaths per produced unit of energy than nuclear power. Coal power doesn't frighten people a lot either, despite the million yearly deaths. Which is worse, the amount of fear or the amount of victims?

With nuclear power, we get a steady supply of energy with puny carbon emissions. But nuclear power has its drawbacks too, and nuclear plants are expensive to build. Should we put our money in them, after all? Or perhaps in developing the electricity stores that are required if we are going to use wind and solar energy on a large scale?

We're in a hurry to stop climate change, and we don't know how long will it take to invent the means to store electricity. What if the development takes so long that the climate will heat uncontrollably in the meantime?

Answering these kinds of big and serious questions requires mathematically exact knowledge and a trust in the force of averages when interpreting statistics.

References

Bruckner, T., Fulton, L., Hertwich, E., McKinnon, A., Perczyk, D., Roy, J., Schaeffer, R., Schlömer, S., Sims, R., Smith, P., & Wiser, R. (2014). Technology-specific cost and performance parameters [annex III]. In *Climate change 2014: Mitigation of climate change 2014* (pp. 1329–1356). Cambridge University Press.

Burke-Hubbard, B. (1998). The world according to wavelets: The story of a mathematical technique in the making.

4

How Mathematics Helps Doctors

Statistics Has Traditionally Been Used as an aid in medicine. It showed its strength in understanding the 1854 London cholera epidemic, when John Snow marked the cholera cases on a map and was able to infer that a polluted well was causing the disease. This is an early example of *spatial* statistics.

The other large service that statistics performs to doctors are tests of medicines with a large number of participants. When a new treatment is tried on a large enough number of patients and a placebo is given to a control group, one gets very reliable knowledge about the effect of the treatment. The bigger the number of the participating patients is, the more certain the result. This is due to the same law of large numbers that waters down the uncertainty due to random variations in the case of rolling dice.

In this chapter I'll tell you how vaccine studies that have been done on a large number of patients lead to practical estimates of the course of diseases in both vaccinated and unvaccinated children. I also take a peek at measuring one's

© The Author(s), under exclusive license to Springer Nature
Switzerland AG 2021
S. Siltanen, *Step into the World of Mathematics*,
https://doi.org/10.1007/978-3-030-73343-8_4

health at home and the *fat lottery*, which is a method of statistical analysis that I and my friends use.

But statistics isn't the only mathematical friend a doctor has. Inverse problems is my own field, and to a large part it concentrates on developing imaging methods and devices to hospitals. I'll explain in this chapter, how X-ray based *CT scans* or *computer tomography* works. It's like a 3-D X-ray vision!

I'll tell you separately about traditional CT scans and about a newer variation that is based on a lower radiation dose. I also go through a bit different imaging technique, the *positron emission tomography*, where the imaging is done with the aid of a radioactive tracer.

4.1   Why Should Children Be Vaccinated?

So-called "vaccine hesitancy" has become more popular recently. It has already caused bad outbreaks of measles in the US and Italy, and in Finland too some areas are in danger of ending up as scenes for life-threatening anti-vax experiments.

As you can see, I'm taking a strong position against anti-vaccination. What's my expertise here? I'm not a doctor, I'm a mathematician.

The reason for my confidence in these matters is based on statistics and probabilities. Take the example of the MPR vaccine, which protects from measles, mumps and rubella. Every medicine, the MPR vaccine too, has both benefits and adverse effects. We only know the probabilities for the benefits and adverse effects, and can't predict them certainly: a vaccine might lead to adverse effects, and not taking it might lead to illness. Which is worse?

Peer-reviewed medical journals have investigated the statistics of MPR vaccinations in an accurate and reliable way. Let's see how we can interpret them.

Alternative 1

Give the MPR Vaccine as a Part of a National Vaccination Program Then almost all children get the vaccine. Although not everyone is vaccinated, a *herd immunity* is formed, and I'll tell you more about it in an infobox. Because of the herd immunity, the virus can't spread, because most children have immunity and there are no sick children to spread the disease. Before long, measles, mumps and rubella thus disappear from the country. This has indeed happened in Finland: there are less than five cases a year, and they have been brought from abroad.

All of the vaccinated children are still at risk of adverse effects! One can illustrate the probabilities with an imagined two-part game of dice.

When the child is vaccinated, the first round starts. It's led by Miss Misfortune, who rolls a die. If the result is a one, then the child will get a mild fever, a pain or a rash.

All of the vaccinated children participate in the second round too. There Miss Missfortune rolls six dice at the same time. If all are ones, then the amount of platelets in the child's blood declines for a couple of months, and this can be seen through increased bruising. If you have been playing Yatzy and have tried to get a one with all of the dice, you know that it doesn't happen often. Even if the child would get this rare result, it would not cause permanent damage.

Alternative 2

Don't Vaccinate the Kids at All   When a kid then gets the measles, the imaginary game is much harsher than in the alternative 1. First Miss Misfortune rolls four dice. If all of the dice give ones, then the child gets an inflammation of the brain. It can lead to permanent brain damage.

This game has a second round too, and every sick person will have to participate in it. There the Grim Reaper himself takes the dice. For a child living in the western countries, he rolls four dice. If all are ones, he takes the small patient with him. In a country suffering from malnutrition, the game is even more gruesome: one die is rolled, and if the score is smaller than a 4, the disease is deadly.

Mumps and rubella have too their own dice games, one even more dreadful than the other. And not only kids play them: rubella can cause miscarriages and foetal damage in pregnant women.

Alternative 3

Vaccinate the Other People's Kids, But Not Mine   Then my child is spared both from the possible adverse effects and the diseases. Although this only happens if a large enough portion of the other people's children are vaccinated to preserve the herd immunity. If the vaccine coverage even of a single kindergarten drops enough, an epidemic could break out and invite both Miss Misfortune and the Grim Reaper to the homes of the unvaccinated kids to lead the Alternative 2 roulette. An unvaccinated person can end up in that life-threatening game when travelling abroad, if he ends up in an area with no herd immunity.

I took the MPR vaccine as an example above, because we have available statistical data of it that is based on a large number of cases. The force of averages then guarantees that the rules in my imagined games correspond to probabilities in real life.

There isn't always a lot of research data on new vaccines, and then the health dice games might have unknown rules. For example, the Pandemrix vaccine was used for the swine flu epidemic of 2009 and it unexpectedly increased the risk of getting narcolepsy.

The rule of thumb is, that the more people have been studied in the vaccine studies, the more certain we can be that we know the risks. Vaccines that have been in use for a long time are safe and help remove plagues like the smallpox.

Box: Herd Immunity

If a sufficiently big part of the population is vaccinated against measles, for example, then a herd immunity forms. Then even most unvaccinated persons are spared of the disease. The formation of a herd immunity can be seen in mathematical models for the spread of infectious diseases, whose core is *exponential growth*.

Exponential population growth
Imagine that ten children were born into all of the families in the world from now on. There are five children for each of the parents, and each of them will get ten children with his or her spouse. The number of people here multiplies in this model according to *exponential growth*, which includes the doubling of the population over a certain fixed amount of time.

What if there was only one child in all families from now on? Then two parents would have only one child, so the number of people would decrease. This is *exponential decrease*, where the number of people is halved during a fixed time period.

(*continued*)

Box   (continued)

These simple models end up like this: if the number of children is greater than a certain threshold, the population increases exponentially. If fewer children are born into families than the number of the threshold, then the number of humans is halved over and over again until mankind goes extinct.

This is of course only an intermediate model for population growth, because the idea that all families have the same number of children forever simplifies things crassly.

Exponential growth in epidemics

The SIR model describes the spread of infectious diseases, and exponential growth in it explains herd immunity. There are three kinds of people in the SIR model. The S group includes the persons who are susceptible to an infection. The I group includes the infected persons and persons who are spreading the disease. In the R group there are vaccinated and recovering people who can't get infected.

The SIR model has an initial state fed into it: the number of people and how many of them belong to each of the three groups. The model needs the daily meeting probability for persons and a number that describes how easily the disease spreads from person to person. The model calculates how the percentages of the groups S, I and R change over time by using *differential equations*. As a result we get a daily report of these ratios as many days to the future as we wish.

The SIR model leads to the exponential spread of a disease or an epidemic, if the disease is infectious enough, people meet often enough and too small part of the population has been vaccinated. If the disease spreads poorly between people, people don't meet a lot and the proportion of vaccinated people is big, the spread of the disease obeys the law of exponential *decrease*. Then no epidemic takes place, but the individual spreader manages to spread it to the few people near him at most and not further. At last the disease disappears completely like for example smallpox.

For more on SIR models, please see the video "Mathematical modelling: How does a virus spread in a population?" on *Samun tiedekanava* in YouTube.

(*continued*)

Box   (continued)

The formation of herd immunity
The alternative predictions of exponential growth or decrease are completely opposite: (1) either a big epidemic or (2) only a few people get ill and the disease disappears. These alternatives correspond to rampant overpopulation and extinction in the population model, whose difference was due to the number of children and its threshold value. In the spread of diseases, vaccine coverage is the corresponding threshold value. If it is high enough, a herd immunity forms and the disease can be contained or even eradicated.

In the case of measles, over 95% of people have to be vaccinated to form a herd immunity, because measles is very contagious. A rubella epidemic can be avoided with 85% vaccine coverage, because it is not as contagious as measles.

The SIR model matches well with practically measured infectiousness factors and epidemics that have taken place. The thresholds it gives for vaccine coverage are approximate, but the existence of such thresholds is mathematically clear.

4.2   The Fat Lottery and Predictions of Weight

I lift kettlebells with a group of friends every Monday. We pump iron hard in the training, but the poor level of banter is even harder on the trainers! The toughest moment of the evening is the playfully named *fat lottery*. The lots are drawn by a scale that measures the makeup of the body.

Here are instructions for the players of the fat lottery: enter your age, height and gender to the scales with a keyboard. Step on the scale barefooted and take a metal stick in your hands. Wait with your hands down until your weight has been measured. Lift the stick to your front with arms straight and let the scale analyse your innermost being. Say your "lottery number" loud to the bookkeeper at the end.

We record the weight, the fat percentage and the muscle percentage to a table. The high point of the evening is when we interpret the graphs that have been made out of the measures to look better, huffing and puffing at the same time. I highlight three observations about mathematical modelling from these conversations.

The first concerns the body weight index. It is calculated like this: the weight in kilos divided by the height in metres squared. For example, my weight index is $102/(1.8)^2 = 31.5$. That's quite high. It tells about the need to lower the amount of fat in the body so that the risk of adult-onset diabetes would get smaller. But the body weight index too is only an intermediate model, and one should be reserved about its results. The Kazakhstani weight lifter Alexandr Zaichikov is exactly the same height as I am and a bit heavier, so his body weight index is bigger than mine. He doesn't however seem to have an extra layer of fat like I do. The body weight index is then only an approximate number, though it's useful as a coarse intermediate model.

As a second matter I take up the *weight trend* that I've programmed. I coded a graph into the spreadsheet for fitting a line into the weight measurements for the last 10 weeks by using the method of *least squares*. It draws a straight line that goes as close to each measurement point as possible. If the line points down, the weight is going down. Members of our middle-aged fellowship hope for this result. However, often the line is going diagonally up. This causes curses, because it means that the recent trend of the weight is upwards.

The trend line has to do with the domain of validity for a mathematical model. If it points down, it will inevitably cross the X-axis sometime in the future. Interpreted strictly, it would mean that the trainer would disappear entirely! In the case of weight increase the trend line predicts reaching

1000 kg sooner or later in the future. That won't of course happen, and the trend will instead have to be understood as a prediction for the near future.

The third insight into modelling came to the fore when we tried to enter the wrong gender to the scale. For example, If I dial the male gender into the scale in addition to my height and age, I get 32.3 as my fat percentage and 30.8 as my muscle percentage. If I introduce myself to the scale as a woman, my fat percentage rises immediately to 37.8 and my muscle percentage falls to 28.2.

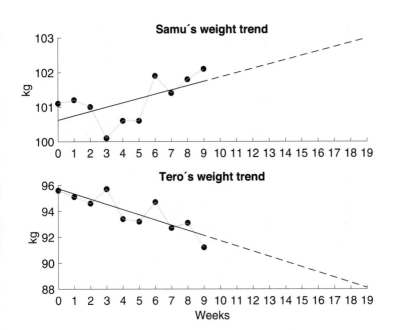

Samu's and Tero's weight in the fat lottery during the last 10 weeks. The scores are the weights in kilos given by the scale. The line is the weight trend that has been fit in to the measurements with the method of least squares.

This is the work of a mathematical model. The scale measures the fat percentage by feeding electric currents between the hands and the feet. The voltages forming in the stick in the hands and the scale under the feet depend on the electrical conductivity of the body. In the terms of school physics is goes like this: resistance = voltage/current. On the other hand, conductivity = 1/resistance, so conductivity = current/voltage. This piece of knowledge can be used to measure the fat percentage, because fat and other tissues have different conductivities.

I once wrote my applied mathematics thesis about *electric impedance tomography,* where electric currents are used to probe the patient's body through electrodes that are placed on the patient's skin. There are usually 32 electrodes. Many *current patterns* are fed and the voltages they cause in the electrodes are recorded during the measurement. The current patterns can be chosen like this, for example: electricity in from electrode 1 and out from electrode 2, next electrodes 2 and 3, then 3 and 4 and so on until the current goes in from electrode 32 and out from electrode 1.

The mathematical task in impedance tomography is to calculate the division of conductivity within the patient based on the measurements. The result is then not one or two numbers, but a picture of the internal organs of the patient. In this way we can follow how the heart pumps blood into the lungs and note if a blood vessel has clogged. You can get more info about impedance tomography on the YouTube video "Sam's Guru Interview: Jennifer Mueller" on *Samun tiedekanava.*

I've researched impedance tomography, so I know well that the measurements on scales can't accurately find out the fat percentage. Four electrodes won't be enough for it at all. How does the scale work then?

Here's my informed guess about how the product development for the scale goes: invite a few 100 volunteer test subjects representing diverse age groups to the factory. Some must have the fat percentage of a marathon runner, others at the top of the scale, and some comfortably at the middle. Divide the group into men and women and then measure their fat percentages with some accurate method, like *hydrostatic weighing* that involves immersion in a bathtub. Collect electric readings of everyone with a prototype scale, which aren't of course enough for determining the fat percentage by themselves. In the end draft a mathematical formula that needs the weight, height, age, gender and the measurement value from the scale as inputs. The scale uses them to measure an approximate fat percentage. The accurate measurement from the bathtub trick are used to calibrate the formula.

One can use the traditional way of least squares to draft the formula. Machine learning is a more modern way, where a mathematical model of nerve cell networks is taught with optimization to "understand", how to generalize the results from the test subjects so that they give the customers reasonable readings.

There's a separate formula for men and women, because the average fat percentage of women is bigger than that of men of the same weight and height. That's why changing the gender on the scales makes a big difference in the results. The scale doesn't then directly measure the fat percentage, but its incomplete measurements are used as inputs in a prearranged formula.

There's a link with the gender models that I discussed earlier: the division into men and women is simple and easy to input into the scale, but what's the right choice with a person with the XYY chromosomes, for example?

4.3   Traditional Tomography

If you have watched *Syke, House* or *Grey's Anatomy* on TV, you have seen tomographic images of patients. This technology is based on X-rays and it's called computer tomography, or a CT scan. A doctor can use it to see the internal organs of the patient without cutting him open on the operating table. A CT scan takes cross-section images of the area of the patient that's being studied.

Tomography works with the aid of mathematical modelling and calculations. It's an example of an *inverse problem* that proceeds from effects to causes. In tomography, the "cause" is the internal makeup of the patient and the "consequence" is how the X-ray photos taken from different angles look.

Ordinary X-Ray Imaging

Let's first think about an ordinary X-ray image. It's based on X-rays that are the same kind of electromagnetic radiation as visible light, but they're a lot more penetrating. X-rays travel along straight lines unlike light, which is easy to bend with glass lenses. The rays travel effortlessly through a human being and they dim, or *attenuate*, according to what types of tissues stand in their way. A ray with bone in its path attenuates more than one travelling through muscle. That's why bones stand out as much whiter areas as the soft tissues that are dark grey.

The special feature of X-ray pictures is the overlap of objects. Think about the situation where the X-ray camera is horizontal like a table, and an X-ray source points straight down. We find two test subjects, and one of them is engaged and the other is not. We then ask them to put their left hands on the top of each other and the detector plate of an X-ray camera, and then take an X-ray picture. There are two skeleton hands and one ring floating in the air in the picture. But it's impossible to tell from the picture, on whose hand the ring is, on the top or the bottom hand! The picture is the same, no matter which way the hands are.

Here we need an official notice: the picturing of hands above should be kept purely as a thought experiment, because X-rays are harmful. The radiation load that is collected in a human body increases the cancer risk. That's why X-ray pictures are taken only after a doctor has judged that the benefit of a picture is bigger than the risk it causes.

On the other hand, the radiation dose from such a picture of one's hand is 0.001 millisieverts or only a *four thousandth* of the dose one gets from natural background radiation in Finland every year. So our test subjects wouldn't end up in great danger because of the hand X-ray.

How Tomography Works

The end product of tomography is a computer model, where one can observe all of the internal structures of the patient without any problems with overlaps. The three-dimensional model can be turned, cut, zoomed in and out. One can make virtual trips in it. Parts of the body, like the jawbone, can be isolated and sent to a 3D printer. Today, tomography is an everyday tool of doctors due to its versatility and usefulness.

One of the most important uses of tomography is identifying strokes or disturbances in the blood flow of the brain. A blood clot or a thrombosis may clog the flow of blood to a region of the brain, and one talks of an infarction. A stroke can also be due to a cerebral haemorrhage. In both cases the external symptoms are the same: a smile on one side of the face, confused speech and not being able to raise one hand above the head. One can treat a stroke with a thrombolysis or a medication that thins the blood, but one must not give it to a cerebral haemorrhage patient. A tomography of the head reveals a haemorrhage, because the haemorrhage can clearly be seen as a white area. On the other hand, a stroke due to a thrombosis or a blood clot is a dark area. In the latter case a doctor can rule out a haemorrhage and start a thrombolysis treatment.

How does the word "*tomos*", Greek for "slice", come into the name of the method?

In a CT device, the patient is lying on a bed that's placed in a big doughnut-shaped machine. There's a rotating ring within the doughnut. A powerful X-ray source has been fixed to the ring, and there's a row of X-ray cameras on the opposite side. The ring spins wildly when images are taken and takes X-ray pictures of the patient from every side.

First, second and third generation CT devices measured a millimetre-thick slice from all sides and calculated a two-dimensional image of it. After that, the bed was moved a millimetre in the doughnut and the next slice was measured. Going like this allows one to make a three-dimensional model of the patient by combining the two-dimensional images.

The idea is the same as when slicing a fruitcake roll into slices: the cross-sections reveal, where the apricots, raisins and canned cherries are located within the cake. An X-ray tomography gives the same information but without breaking the cake.

Fourth generation CT devices have moved on to spiral geometry, where a two-dimensional camera takes an X-ray video as the patient slides through the device. The term tomography is here misleading, but it's still used for historical reasons.

How to Calculate the Tomographic Model from the X-Rays?

Filtered back-projection is the most often used method. It goes through every X-ray the device measures like this. Mark the place in the picture, where an X-ray travelling along a straight line ends up. Choose out of the measurement the attenuation value for the X-ray. The attenuating value gives information, how much attenuating tissue the ray faced when passing through the patient. A lot of attenuation means a big number, a little of attenuation is a small number and a ray passing through air gives a zero. Put the number into all pixels that are on the path of the X-ray.

The entire collection of the measured X-rays is gone through as above, and the values for each pixel are added together. This operation is called a *back projection*.

After the back projection has been finished, attach different grayscales to the numbers in the pixels. A zero is still black and the biggest number in the picture is white. The numbers in between are given a grey colour on an even scale in proportion to the numeric attenuation value of the pixel. The picture that is formed in this way is quite good, but blurry.

This blurriness can be removed by emphasis on the lines, which is called a *high pass filter* in professional circles. Technically it is done by moving into the frequency domain by using a transform that was developed in the nineteenth century by the French mathematician Joseph Fourier. One can emphasize higher frequencies in the frequency domain so, that the inverse transform gives a sharp and accurate tomography of the patient.

A History of Tomography

The history of the filtered back projection is interesting. The first one to publish on it was the Austrian mathematician Johann Radon in 1917. Radon's study was pure mathematics and didn't have anything to do with practical applications. So his formula was for a long time only of interest to theoretical mathematicians.

Allan McLeod Cormack developed corresponding solutions in the 1950s and published them in the early 1960s. He didn't know of Radon's earlier work. Cormack was already thinking of tomography, but he didn't realize it in practice.

Godfrey Hounsfield was working in England in a company called EMI at the same time, and he started to build a tomography machine. The financial situation of EMI was great, because the Beatles were selling a lot of records. That's why the company could afford to keep an eccentric inventor building X-ray devices on the payroll. Hounsfield was inspired by Cormack's work, although he didn't use a filtered back projection to calculate the tomographic image in his first 1971 device (he used a pixel-based system of equations instead).

There's an interesting detail about Hounsfield's early images. He started with a tomography of the head and was able to see the contents of his own skull, among other things. His picture showed a dark dash between the skull and the brain. Hounsfield inferred that there's an air gap between the brain and the skull. This astonished doctors, because the air gap had not been observed in autopsies or in other conditions either. It was later revealed to be a mistake that was due to Hounsfield's way of interpreting the X-ray measurements. He used a simplification that assumed all of the X-rays to be of the same colour. (The X-rays have their own wavelengths or a "colour", like visible light has and the rainbow shows. We just can't see X-rays or their colours.) In fact, a collection of different colours comes from an X-ray source. The problem can be fixed by taking the colours into account when calculating the tomographic image, and the "air gap" between the skull and the brain disappears.

Cormack and Hounsfield got the Nobel Prize for Medicine in 1979 for developing tomography. Their work began the development of CT machines that has been hectic. Most devices use corollaries of Johann Radon's formulae to make very accurate three-dimensional images of patients. Sometimes I wonder why it took 60 years from the mathematical invention made in 1917, before the useful formula ended up serving mankind. In part this is due to the development of computers that allow one to calculate tomographic images in practice. Another partial reason is that mathematicians easily immerse themselves in their theoretical worlds and don't start trying out their ideas in practice.

I encourage mathematicians to work together with engineers and everybody else to talk with mathematicians about their ideas and their potential applications!

You can get more information about tomography on YouTube on *Samun tiedekanava*. Search for a video called "What is tomography?"

4.4 The New Low-Radiation Tomography

Traditional tomography is used widely and it works well. Do we mathematicians have anything left to study on this subject?

In fact, there's a lot to do in reducing the radiation doses. For example, a CT scan of the hip causes a patient a 10 millisievert radiation dose, which corresponds to the dose one gets from natural background radiation during two and a half years. Therefore, one shouldn't do many such scans on a patient during each year.

Would it be possible to do a tomography with a lot less X-ray radiation?

In my research group we've noticed that it can be done, on certain conditions. The mathematics needed to compute the picture is more complicated than in the case of imaging with a full radiation dose, and the result is always imperfect. That's why a low dose CT scans are always developed with a medical need in mind. The method for imaging is developed to ensure that the final tomographic image is good enough for this need despite its deficiencies.

A CT device for designing dental implants was designed in Finland. It's is a good example of this kind of low-dose tailor-made CT imaging. I'll tell you more about it below.

An Inverse Problem and Two Penalties

Solving an inverse problem starts with understanding and mathematically modelling the forward problem. A forward problem means moving from the cause to the effect. In the case of tomography, it would mean finding out the X-ray image taken from a given direction, if the internal makeup of the patient is fully known.

The forward problem in tomography is solved by dividing the patient into small imaginary cubes of *volume elements* or *voxels*. Each voxel is thought to have only one type of tissue. If the voxels are chosen to be small enough, we can represent all internal organs accurately enough with them. This matter about accuracy is a bit like pixel resolution in digital photography: more megapixels leads to more detail in the picture.

The model follows each X-ray travelling along a straight line from the point source in the X-ray tube to the pixel in the radiation detector. The model then takes the distance the ray travels in each voxel and its type of tissue into account. It's as if the computer was taking an imaginary X-ray of a patient whose internal makeup had been written as numbers into voxels.

Box: Pixels and Voxels

If you zoom close enough to a digital picture, you get to see square-shaped picture elements containing only one colour. The picture elements are also called *pixels*, which is inspired by their English name. One cannot see details that are smaller than a pixel in the picture, whereas the details in an oil painting can be as small as a molecule, for example.

Picture elements are used in the electronic picture formats, because a computer cannot keep a book of innumerable amounts of molecules. It uses a finite list of numbered colours and limits the number of pixels so that they fit in the computer memory.

The term *voxel* comes from the term *volume element*. It's a cube instead of being a square, and piling up voxels can give a suitable computer format for three-dimensional structures. For example, the result of a tomography is a man-shaped pile of small cubes or voxels, each of which has a number that describes how much it dampens X-rays. The number is big for bone, and a voxel in the thighbone is therefore white. The number for the gristly soft tissue in the outer ear is on the other hand small, and therefore the ear is made of dark grey voxels in the CT image.

Solving an inverse problem is based on looking for the smallest penalty with a computational *method of optimization*. The penalty is sentenced on a group of voxels that is filled with numbers, and it is given for two kinds of offences.

The first offence is making wrong kinds of X-ray images. When some voxels full of certain numbers are brought before the court in computer memory, the X-ray model calculates virtual X-ray photos in the same directions as the patient was pictured. The resulting virtual pictures are compared with real pictures, which are already stored numerically in digital form. The corresponding pixel values are subtracted, and a square root is slapped on them. (This is Pythagoras' formula that is familiar in school, but in a higher dimension.)

Comparing virtual and real X-ray photos can be described briefly: if the virtual images resemble the ones taken from the patient, then the fine is small. If the virtual images look very different than those taken from the patient, then the penalty is big.

The first penalty isn't enough alone to find the right 3D model of the internal organs of the patient. Because we're trying to minimize the radiation dose by taking only few images, very different patients might end up with quite similar pictures. In these cases, calculating a 3-D model of the patient is prone to measurement errors. There are always small errors in measurements, because the position of the device is never known infinitely accurately. In addition, the pictures taken by the X-ray camera contain noise that looks like snow, much like ordinary photos that are taken in dim light.

Therefore, a second penalty is needed, which is sentenced on sets of numbers in voxels that are not like human tissue.

For example, negative numbers aren't allowed, because then the X-rays in these voxels would be strengthened rather than weakened. Voxels standing next to each other shouldn't have very different numbers, except in the boundaries of different organs and types of tissue. These boundaries are however quite rare when compared with the number of voxels inside the tissues.

The computer then faces the following situation: certain numerical values are proposed to the voxels, and it must calculate the corresponding penalty. There are really two penalties, and they are added together. The first penalty is the bigger, the worse the virtual X-ray images correspond to the really measured X-ray images. The second penalty is the bigger, the less the numerical values of the voxels correspond to ones found in the human body.

Using Optimization to Find
the Smallest Penalty

The values of voxels giving the smallest total penalty are calculated using optimization. The calculation proceeds step by step, improving on the best estimate of the moment.

Having zero as the value for all voxels is fed into the program as the first guess. Then it's clear that the penalty is big, because all of the virtual X-rays of that stack of voxels are completely black and don't therefore look at all like the image taken of the patient. Next the method should find new numbers for the voxels so that the penalty is smaller. This is done with optimization.

Box: Optimization

The goal of optimization is to find the best possible circumstances. For example, a family apartment should be big enough so that there's enough room for everybody. On the other hand, every extra square metre costs money, and a big home is more difficult to clean up. Differing house prices per square metre and different difficulties in making the trip to the grocery store add extra problems to the optimization of apartments. Optimization involves seeking for the best combination of all these different considerations.

In the world of mathematics, the optimization task must be well defined with precise numerical *penalties*. Designing a penalty may take a lot of work. It could depend on the place, like house prices per square metre and the trip to the grocery store show. But how to give a penalty to a flat that's too small? Of course, a study of 10 m² is worse for a family of five than a mansion of 120 m², but which exact number should one add to these alternatives? Designing these kinds of penalties is a central part of mathematical modelling.

How are optimization problems solved in practice? Imagine an explorer surprised by a fog in an icy mountainous landscape. He should find his way back to the bottom of the valley, so I'm choosing the altitude as the penalty. The peak

---

(*continued*)

Box   (continued)

is the worst place to be, because there the penalty is the greatest. The destination or the lowest point in the valley is the place that gives the least possible penalty. Optimization looks for the place with the lowest penalty, and the explorer succeeds when he reaches the bottom of the valley.

Which principle should the explorer use when looking for a way down? Due to the fog he sees only a metre to the front, so he doesn't know in which direction the destination is. The *method of steepest descent* invented by Augustin Louis Cauchy in 1847 comes to the aid.

The method of steepest descent resembles sliding an icy hill down on skates. Our explorer puts a pair of them on his feet. He chooses a direction, where the hill seems to descend the steepest for the visible 1 m, and starts going downhill in that direction. Due to the length of the skates, the descent looks straight from a bird's eye view. When the downhill slope ends and turns to an uphill slope, the explorer stops. He is now lower than before, but not necessarily at the bottom of the valley. He turns his skates again in the direction of the steepest downhill slope and makes another descent along a straight line until the route starts to rise again. This way he provably gets to the lowest spot of the closest valley.

Cauchy's method is stylish but slow. It has been improved in many ways over the last decades. Some of them make a cleverer choice for the direction of the descent and others fix the length of the landing with a different rule than waiting for the descent to turn into an uphill slope. Probabilistic models parachute explorers from helicopters all across the landscape and choose the skater in the fog who has got the furthest down at the end of the day.

The downhill skating example represents optimization in a case with two dimensions, because the position of a skater can be given with two coordinates when looked from the above: a north-south coordinate and an east-west one. The method of the steepest descent and its more efficient variants also work in solving higher-dimension optimization problems. Our apartment example is one of these, because already the position of the house already includes two dimensions. In addition there is the floor area of the mansion, its age and "dimensions" that are more difficult to grasp, like the layout of the rooms and whether the house has been renovated.

Box: Higher Dimensional Spaces

The optimizations required by tomography take place in higher-dimensional spaces. Each voxel represents its own dimension, and there are millions of voxels in our model. Minimizing the penalty is a hard task even for the computer in such a complex geography.

But how can there be more than three dimensions?

Normally it's enough for us to tell our location on Earth with GPS coordinates, so that the longitude or east-west position is the X coordinate and the latitude or the north-south position is the Y coordinate. Altitude from sea level is then the Z coordinate. A negative value of Z is depth under the ocean. This is a three-dimensional space that is familiar to us.

We can certainly take time as a fourth coordinate by following Albert Einstein's train of thought. This is extremely useful when modelling the structures of the universe, for example. Black holes were first invented by using four-dimensional theory and then observed later on.

But how could there be more than four dimensions?

Let's think of a football fan who follows the Finnish Premier League Veikkausliiga enthusiastically. During the season, he marks every goal by every team into his table. There are 12 teams in the league, so there are 12 numbers in his table, and the numbers are updated after games. A statistical table for the goals for ten seasons is formed like this: The numbers of the first column are added to get the goals of FC Honka over the whole period of time. One gets the second number by adding all FC Inter goals. Each team has its own column in the table, and adding the goals of a team is independent of the goals scored by the other teams.

In the same way, an explorer can go directly south so that the Y coordinate changes but the X coordinate stays the same. He could also turn 90 degrees and go to the east for some distance, and then the Y coordinate doesn't change although the X coordinate changes. This is the idea of dimensions: the coordinate for each dimension can change without the other coordinates changing.

It's difficult for a human being to visualize more than three dimensions. But there are no mathematical difficulties in adding dimensions: we only need new coordinates or columns for numbers in a table. Travel on the ground takes

(continued)

A Low-Radiation Dental X-Ray Device

Has your dentist taken images of you with a panoramic dental X-ray? That's an X-ray gadget that goes round the head and takes a big "smiley picture" of all of your teeth? You've probably been to one of those, because a panoramic X-ray has been a basic tool for dentists since the 1970s.

A panoramic X-ray gives a good overview of teeth and their roots, the sinuses, the ends of the jawbone and of many other bone structures in the head. The odontologist Yrjö Paatero developed this imaging technique when he was working at the University of Helsinki in the 1950s. The efforts of the Finnish Palomex corporation made panoramic X-ray devices, or *orthopantomographs*, a standard piece of equipment in dental clinics across the world.

In the early 2000s, I worked in a product development team at Instrumentarium Imaging, and our task was to do tomography with an orthopantomograph. We were especially thinking about helping dentists in installing dental implants. To install an implant, one has to drill a hole in the bone where the missing tooth was, and place a screw there to support the implant. One must not drill the hole too deep, because there is a risk of damaging the nerves, like in the case of lower jaw. Usually the right depth is looked up from an accurate 3-D tomography that is taken with a cone beam CT machine. But our problem was to develop a corresponding solution with a smaller radiation dose.

We programmed new movement paths into the panoramic imaging device, so that we could take X-ray images of the patient's jaw from different angles. There were some difficulties, because the panoramic device is designed for a totally different task and it couldn't move the best way possible. In the end we were satisfied with taking 11 X-ray photos and limiting to an angle of 40 degrees. We weren't able to take pictures from all around the patient as in traditional tomography, but only within an angle of 40 degrees. This leads into an even more difficult mathematical puzzle when calculating the tomographic image.

We were able to develop a new imaging device to the market after years of work and with a big development team. Its three-dimensional reconstructions aren't as fine and accurate than ones made with a cone beam device, but they're good enough to determine the safe drilling depth for a dental implant. From a dentist's point of view this is a good solution, because it's not necessary to buy a new expensive imaging device to the clinic. The clinic's panoramic device can produce tomographic images to help in implant planning after a mathematical computer software upgrade.

From the point of the view it's especially nice that the new gadget we call the VT device gives you only one hundredth of the radiation dose as a cone beam tomography causes.

There's a video on the VT device on YouTube on Samun tiedekanava. It's called "Sam's Science Splash: Low-Dose Dental X-ray Imaging."

A Tomography that Uses Radiotracers

We've just described X-ray tomography. Its idea is to place an X-ray source on the one side of the patient and an X-ray camera on the other, and then use the dimming of the X-rays to measure the thickness of the tissues that stand in the way of the X-rays. Although forming the image is a challenging mathematical problem, it's made easier by the fact that we know the place of the X-ray source and the camera pixels.

In isotopic imaging like *positron emission tomography* or a PET scan, the radiation source is inside the body. A small amount of radioactive tracer is placed in the bloodstream, and one can follow its location in the body with external measurements. One a common tracer is fluorodeoxyglucose, which is formed when a radioactive fluorine atom has been artificially placed in a sugar molecule. To put it more accurately, an OH group in a sugar molecule is replaced with a fluorine atom, whose isotope is 18.

The body still interprets the modified molecule to be sugar, so it is transported to a part of the body that needs energy. But the fluorine is not a part of sugar and prevents its use as a source of energy. Therefore, parts of the body that are thirsting for energy, like cancer tumours or parts of the brain that are thinking actively, end up as concentrations of radioactive fluorine.

When fluorine-18 splits, a positron darts out of the nucleus. It's antimatter, and it inevitably collides with an electron and creates pure energy. Usually this takes place after a flight less than a millimetre. The pure energy turns into two powerful X-rays that fly in exactly opposite directions from the point where the electron and the positron collided. The X-ray cameras outside the body have been set to notice simultaneous flashes. So when two pixels send a signal at the same time, we know that an electron and a positron have met on a line between the two pixels. That in turn tells that there are fluorine-18 isotopes near the meeting spot and thus active metabolism takes a lot of energy there.

The mathematical image formation for PET scans is formed like the calculations with voxels and penalty functions that I described above. In the case of PET scans the situation is more complicated, because we don't know where the radiation comes from and its source is a part of the inverse problem.

You can get more information about PET scans on Samun tiedekanava on YouTube. See the video "Sam's Science Splash: Brain Imaging Using Antimatter."

Box: The Isotopes of Fluorine

What's the difference between the fluorine isotopes 18 and 19? And what are isotopes at all?

Matter is made of molecules, and their parts are atoms of chemical elements. There's only a limited number of elements, and they have been listed in the periodic system. An atom is made of a nucleus with protons and neutrons, and a shroud of electrons. The number of protons in the nucleus tells what element it is. A proton has a positive electric charge, and the electron shroud has as many (negatively charged) electrons as there are protons in the nucleus.

Neutrons don't have an electric charge, and there could be different numbers of them in the nucleus even though the element stays the same. Atoms of the same element with differing number of neutrons are called the isotopes of the element.

Fluorine has many isotopes with 5–22 neutrons in the nucleus.

(continued)

Box (continued)

Ordinary non-radioactive fluorine is of model fluorine-19. There are 9 protons and 10 neutrons in its nucleus. It's a comfortably balanced combination that holds together practically forever. We use it to brush our teeth and smile afterwards.

Fluorine-18 is the most interesting isotope from the rest for imaging purposes. Its nucleus has 9 protons and 9 neutrons. The combination is unstable so that an atom can move to a lower energy level through radioactive decay. Usually a so called beta decay takes place. Then one of the protons is turned into a neutron, spitting out one positron and one neutrino. An atom with 8 protons and 10 neutrons is left, and that's the stable isotope oxygen-18!

The laws of nuclear physics are such that we can't predict the moment a single nucleus of fluorine-18 decays. But we do know that if we have a given amount of fluorine-18 in a can, a half of it will have been transformed into oxygen after the next 2 h. After the 2 h after that, the amount of fluorine is quarter of what it was at the beginning, and the halving just goes on in periods of 2 h.

And what about the decay products? The neutrino is kind of a "ghost particle" that just flies away without interacting with anything. But the matter is different with the positron: it's antimatter that is transformed into pure energy when it collides with ordinary matter. Medical isotopic imaging uses this property.

5

Is Humanity Too Just Mathematics?

Successfully describing deeply human things with mathematics requires careful work. Take the example of a conversation between a computer and a person. The computer plans its answers with formulae for calculation, and it's usually easy to tell the result from an exchange of thoughts between two people. Or what do you think about my chat with an Internet chatbot?

Samu: Do you like ice cream?
Chatbot: I don't understand. I might do a better job if you give simple keywords.
Samu: OK. Where are my house keys?
Chatbot: I ran a search in our help centre. Read the article "Keyboard shortcuts."

The mathematical instructions of that chatbot don't imitate human conversations very well. Of course, I was unfair too. The bot I mentioned wasn't designed as a general conversation partner, but to help with a limited range of technical problems.

© The Author(s), under exclusive license to Springer Nature
Switzerland AG 2021
S. Siltanen, *Step into the World of Mathematics*,
https://doi.org/10.1007/978-3-030-73343-8_5

On the other hand, computers have been able to reach unbelievable feats in tasks that have earlier been believed to be a human's job. For example, a human can't beat a computer at chess or at go anymore. A computer can today recognize cat photos, even though that was unimaginable even a few decades ago. I tell you more about this in this chapter.

What about culture? At least I like to think that my taste in movies, books and music are very fine, unique and refined. A computer couldn't make sense of them. Or could it? And could an AI create art?

5.1   A Poor Model Produces Poor Results

A mathematical model is always just an intermediate model, so it inevitably fails to describe its object. These deficiencies often come to light in tragicomic ways when modelling human life.

Let's first examine a crazy model for human happiness:

Happiness = monthly wage + the number of children.

Imagine the ordinary Finn Liisa, who has two kids. Suppose that her monthly wage is the average income in Finland in 2017, 2283 euros. How much will Liisa's happiness change if she has a third child? During her pregnancy the amount of happiness was $2283 + 2 = 2285$ and after the child is born it is $2283 + 3 = 2286$. A new baby raises Liisa's happiness less than half a permille.

If Liisa happened to be the President of Finland and had a salary of 10,500 euros a month, the baby would have increased her happiness with less than a hundredth of a percent. Why would a person with a high income be less happy when a child is born than a person with a low income? How many people would be made happier by a two-euro raise than the birth of a child? Moreover life is complex, and the birth of a child doesn't increase happiness always and in all cases. This model doesn't describe reality in the least.

Translating literature has been a field of human activity that has been a no-go for computers. I illustrate this with an example by taking the first the first paragraph from Veijo Meri's *Peiliin piirretty nainen* (1963, translator LS):

> *A bus from the Kiito Line stopped next to store on the other side of the road. The driver waited so that a few smaller cars had time to drive away. Then the bus came across the street and when it rose to the sidewalk, it swung lightly from side to side as if it were trying to walk. A curb stone sank deep into its rubber tires.*

I use Google Translate to translate the Finnish version into Swedish and then back into Finnish. The result is this:

> *The Kiito Line bus stopped in a store across the road. The driver waited for the smaller cars so that he could drive away. Then the bus came across the street and as it became the sidewalk it turned slightly from side to side as if trying to go. The boundary of the lamb sank deep into the teeth gums.*

One could charitably characterize this as a new text with poetic merit. But the twice translated paragraph doesn't fit the opening of Veijo Meri's realist-styled novel at all.

These examples show that poorly designed mathematical computations cannot catch the nuances of human existence and action. What about better models?

A lot of scientific models have been proposed for happiness, and some have even been tested by comparing them with data from magnetic imaging scans of the brain. The best models are a lot more complex than the above one and can reach very interesting results. I won't get deeper into them here.

Machine translations of literary texts has improved greatly during the last few years. Translating the first paragraph of Ernest Hemingway's short story *The Snows of Kilimanjaro* from English to Japanese and back succeeds remarkably well (Lewis-Kraus, 2016):

Version 1   Kilimanjaro is a snow-covered mountain 19,710 feet high, and is said to be the highest mountain in Africa. Its western summit is called the Masai "Ngaje Ngai," the House of God. Close to the western summit there is the dried and frozen carcass of a leopard. No one has explained what the leopard was seeking at that altitude.

Version 2   Kilimanjaro is a mountain of 19,710 feet covered with snow and is said to be the highest mountain in Africa. The summit of the west is called "Ngaje Ngai" in Masai, the house of God. Near the top of the west there is a dry and frozen dead body of leopard. No one has ever explained what leopard wanted at that altitude.

It's difficult to say from the texts, which is the original and which has been twice translated by a computer! The creation of the computer has changed during the translations, so it's not the same word for word. But it's a beautiful text and small editing would allow it to reach the standard of literature.

But why didn't the double Veijo Meri translation succeed nearly as well? Both have been translated with the same Google Translator. This is due to neural network modelling.

The quality of Google translations leaped upwards when Google moved to using neural networks, or artificial intelligences that coarsely imitate the function of the brain, in its translations. It made the fantastic Hemingway result possible.

Machine learning for translations is based on analysing a bilingual text archive on neural network models. A computer is offered a large number of texts which tell the same thing in English and in Japanese. During the learning process, the computer adjusts the internal functioning of the neural network so that the given texts are translated into each other as accurately as possible. The more material is available, the better the learning result.

The linguistic areas for English and Japanese, and the cultural exchanges and commercial messaging between the US and Japan are so big, that there are vast amounts of textual masses in the digital archive. That's why a neural network learns to translate well from English to Japanese and vice versa.

Maybe there aren't enough comparable texts between Finnish and Swedish to raise the level of translation from that level of "poetic" Veijo Meri translations. Of course, it could be that our small language areas arent on the top of Google's list of translation development priorities.

I'll next tell more precisely about the functioning of neural networks.

5.2 The High Points of AI

A Self-Taught Chess Machine

Neural network-based machine learning has reached astonishing feats during the last 10 years. These coarse models for brain activity have made computers to perform well in tasks that in the past were thought to be possible only for humans. They include speech recognition and translating texts, which was discussed above. AI is also intruding into art; I'll tell more about it in the next section.

I was quite a math nerd as a kid. I played some chess like math nerds do, often with friends and family. I read some beginners' guides and solved chess puzzles. I never dove very deep into this world, and I'm not much of a player.

My greatest chess achievement was in a match against my schoolmate Teemu Keskisarja in the 1990s. Teemu was then one of the best players in Finland. At first, Teemu responded to my moves fast with the routine of a tournament player, but soon my strategic knight jump gave him quite a pause. Did I catch the master? Teemu's moment of thought is without doubt the high point of my chess career.

But it wasn't about catching Teemu in a trap. My move was so nonsensical that Teemu had to think whether there's a devilishly clever plan behind it. So I suffered a quick and crushing defeat!

That's chess. An enlightened amateur has no chance against a national-level top contender, who in turn can't score points in tournament tables for international grandmasters. Chess requires hard logical thought to assess the next few moves like this: "If I do like this, then he can do like that and these relative strengths follow from the moves." However, nobody can calculate forward more than the next few moves, because there are so many possibilities. Therefore good players have the experience to make the moves that support their long-term strategy, strengthen their positions and help their future attacks, no matter for the details of the attacks.

For a long time it was thought that the intuition of an experienced champion is an unbeatable advantage over the raw calculating power of a computer. But the IBM chess computer Deep Blue beat the world chess champion Garry Kasparov, and after that the triumph of computers over human players has just gotten faster. A chess computer can calculate millions of moves each second and choose the move leading to a better position from there on. There's also a huge collection of games in the computer memory, and the computer can learn from them.

A new turn in chess history took place in 2017, when the AlphaZero AI program beat Stockfish 8 that was based on traditional calculating techniques. AlphaZero does not have archives of games and it doesn't calculate moves to the future. It started to play against itself and learnt or *machine learnt* from its games. In 4 h, it learned to play better than Stockfish 8, which is a result of decades of programming work.

AlphaZero uses *neural networks* for machine learning, and the networks are mathematical models for the functioning of the brain (Silver et al., 2017). I'll next tell you how they work. Don't worry if you don't know chess: there are no fine points of chess rules or tactics in my account. I've learnt from playing against Teemu that I'm not the right person to write them. If you have deeper knowledge of the inner life of the AlphaZero on the other hand, you'll notice the simplifications I've made.

Imagine two square-shaped office blocks. They're exactly alike, except that one of them is white and the other one is black. The plan for the first, second and third floors is like a chessboard: 64 small rooms arranged into a 8 × 8 box. A bureaucrat sits in each office, and his job is to transmit numbers to the top floor according to the instructions that come from top management.

The activity starts when the bureaucrats at the first floor of the white office building is each given a number. The 8 × 8 number grid describes the situation on the board: a zero means that there's no piece in that square. Each piece (pawns, rooks, knights, bishops, queens and kings) have their on numbers, which also depend on the colour of the piece.

The next step is communication from the first to the second floor. Every bureaucrat on the second floor calls nine of the colleagues below: the one right underneath and the eight bureaucrats neighbouring the one below (there are less neighbours at the edge of the board). He asks everybody their numbers and starts a calculation that is described in the firm's *Second floor instructions*. For example, it could go like this: "Take the number of the room right below and subtract the numbers of its northern and western neighbours. Multiply the numbers of the north-eastern and south-western by three and add them to the result of the subtraction. Add a half of the number of the southern neighbour." In the end, the bureaucrat checks if the number is negative. If it is, he replaces the result with a zero.

The bureaucrats in the third floor have *Instructions for the third floor*, which could be like this: "Multiply the number straight below with seven and add the numbers of the eastern, south-eastern and north-eastern neighbours multiplied with 100. Multiply the number of the north-western neighbour by 3.5 and subtract it from the previous result. If the result is negative, replace it with a zero. If the result is bigger than 14, transmit the number 14 upstairs.

The floor plans for the fourth and fifth floors differ from the lower floors. They have only 16 rooms laid out in a 4×4 grid. There are four third-floor rooms under each fourth-floor room.

The bureaucrats on the fourth floor have a simple task: collect the numbers from the bureaucrats working under you and transmit only the biggest of them upstairs.

One acts in the fifth floor like on floors 2 and 3: use the *Instructions for the fifth floor* to calculate a composition of the numbers of the office on the floor below and its neighbours, and send it upstairs.

On the sixth floor, there are only four rooms, where the bureaucrats pick the biggest number from the four fifth-floor rooms under them and send it upstairs.

On the seventh floor there's only one big room where the board of directors works. It takes the numbers from the sixth floor and uses the *Instructions for the board of directors* to compute four new numbers: the row and column of the piece to move, and the row and column of the square where the piece will move into.

The white piece is moved on the board according to the result, and the new position is fed into the first floor of the Black office block. A chain of events corresponding to the one above begins there.

Let's introduce a complicating feature to our imagined situation. Instead of just one office block, there are hundreds of office buildings like the ones described on both the White and Black sides. Each has its own sets of instructions, and some bureaucrats have the straight telephone numbers of colleagues sitting in a neighbouring office block. The numbers from neighbouring blocks are too taken into the calculations following official instructions strictly. The board of directors gets the result from each office block and combines them according to their own rulesets.

How does this kind of calculating lead to a victorious game of chess? It might sound at first that the moves defined like this won't even be allowed by the rules of chess.

In the beginning, they certainly won't be. The neural network model contains machine learning as an important part, and it leads to improving results. Machine learning is done step by step. If the board of directors suggests a bad or a forbidden move, the instructions of the company are modified a bit so that the position leads to a better move. These modifications take place through optimizations, just like in medical tomography too. In machine learning, optimization modifies the coefficients in the bureaucrats' instruction sheets instead of grayscales.

After the program has played a game of chess against itself, both the winning and losing regions of office blocks use the experience from the game to improve its activities. After 1000 of updates, the quality of instruction sheets starts to be so good that the board of directors delivers stronger and stronger moves.

During 9 h of championship training, AlphaZero played 44 million games against itself. Already after 4 h it was so good that other chess programs weren't a match for it any more, not to speak of human players.

The village of office blocks I presented above, with its chains of command, instructions and calculations seems completely insane. Of course it's just a wild simplification of the AlphaZero program. But the core of my presentation is that the real AlphaZero uses these kinds of simple rules at the bottom. There are just very many virtual bureaucrats!

Box: Neural Networks

There are about a 100 milliard nerve cells in the human brain. They communicate with each other by sending electric impulses to each other via axons. The messages pass from cell to cell through *synapses*. There are over 1000 billion synapses in the body, so the network formed by our nerve cells is very complex.

Every nerve cell has a way for handling electric messages. Usually they do like this: a few incoming pulses don't do anything, but if there's an incoming message reaching a certain threshold, the cell starts to send outward pulses too. The more messages come in, the more are sent.

We know that growth and weakening of connections between nerve cells is connected with memory and learning, but the functioning of the brain isn't yet understood on the whole.

The neural network programs used in machine learning are coarse models of the brain. I use an imaginary office blocks with bureaucrats sitting and making calculations, when I'm describing a machine that learns chess. These calculations and their official instructions imitate the behaviour of nerve cells, but their coarseness makes them only intermediate models.

Despite their simplifications, neural network models can reach tasks that were previously thought to be a domain only for humans. The mathematical motor of machine learning is optimization in high-dimensional spaces. The optimization modifies the numbers on the official instructions for the tower blocks, just like the communications between nerve cells strengthen or weaken when a person learns.

A Cat Recognition Program

The office block model for the chess example works in other surprising contexts too. A computer has learned to recognize photos with cats by using a similar model. A child may know how to do that, and you're right. But let's think about the task the computer is faced when interpreting pictures.

A digital photo isn't made out of colours but of numbers when it's in a computer memory. Or numbers that correspond to colours. Just like in colouring books, where the numbered areas must be coloured with the right colour in order to see the picture.

In the case of digital pictures, the areas to be coloured are small squares that are called *pixels*. You've probably seen pixels when magnifying a picture on a computer or smartphone screen. There are three numbers in computer memory for each pixel, and they correspond to the amount of red, green and blue colour in each small box. It's called the RGB colour model for *red, green* and *blue.* Usually the smallest possible number is zero, which means that the pixel doesn't have any of that colour. The biggest possible number is 255. For example, all of the values are zero for black. A bright red is born when R is 255 and G and B are zero. There are 16,777,216 different colours in the RGB model.

When we're looking at the picture on the screen, the fine electronics of our device picks all of the numbers from the file for the digital picture and chooses the right hue for each spot on the screen from the palette of 16,777,216 alternatives. The visual system of our eyes and brains takes care of the rest, and like the child we realize if it's a cat or something else.

What about the computer? Think for example of a picture, where the pixels have been arranged into 1024 rows and 1024 columns, as if in a vast spreadsheet. There are three of these spreadsheets, one for each colour. When the computer should decide whether there's a cat in the picture or not, it must reason by using the numbers in the file and mathematical calculations. The computer has no eyes, no visual system and no childhood experiences, where the little Mörri is purring in the lap.

What kind of mathematical calculation will reveal the feline in the jungle of numbers? The task was unattainably hard only 10 years ago. No wonder: there are cats of different colour, tabby cats, cats of even colour, cats with spots, cats yawning and licking their lips, jumping, sleeping and preying cats. The lighting could be sun at noon, a cellar light or a log fire. The number of possible things that can be seen in the background is difficult even to imagine of. The problem of cat recognition is complex, because the confused nature of reality affects it unlike the imaginary and rule-governed game world of chess problems. Still neural network-based machine learning can solve the cat puzzle.

It's clear that one must be able to dig higher-order features like ears, tails and combinations of cat faces out of the mash of numbers at the pixel level. It turns that the office block and its bureaucrats that I described in the chess example can do the job. It goes like this.

In the first and second floors of the office, one needs a room for each pixel. A bureaucrat sits in each room, in the first floor with the pixel numbers and in the second floor with *Instructions for the second floor.* The instruction again consists of a formula that combines the numbers offered by the nine closest lower-floor neighbours with suitable additions and multiplications.

The instruction on the second floor leads to a big number, if the 3 × 3 grid of pixels downstairs contains a straight line. As in if the room below and its neighbours to the north and the south contain zeroes or black pixels, for example. In addition, the other squares have only numbers 255 for all colours, that is white. In these kinds of cases the bureaucrat on the second floor gets a big number to report upstairs, and a small number in other cases.

The successful identification of cats requires many office blocks. The goal is to build higher-order combinations of the numbers in the pixel jungle. Think about a small circle in the picture, like the outlines of a cat eye. We can think of the circle as approximately consisting of eight short straight lines that are arranged like a stop sign. In a puzzle, the stop sign could be made of nine pieces with the one at the centre reading "STOP". To the left and to the right there are pieces with a direct vertical line or a north-south line. Above and below it there are pieces with a line going left to right, and the corner pieces have slanted lines that complete the eight sides of the stop sign.

So we build three office blocks alongside the mentioned north-south line department: the departments for straight east-west lines, southwest-northeast lines and northwest-southeast lines.

Our cat recognition company continues its construction work. The *House of Circles* is formed. On its first floor, watchful bureaucrats receive information from their comrades who sit in corresponding rooms in the line departments. In the second floor a new number is put together by using the *Instruction for the second floor of the House of Circles*. The instruction works so that a big number gets reported upwards, if the lines below form an octagon like a stop sign. In other cases, a small number is notified upwards.

In the third floor of the House of Circles, there's a compilation floor like one that was seen in the chess example. There are rooms corresponding to four bureaucrats on the second floor. The biggest number from the rooms below is reported up from the compilation floor. Here the idea is flexibility: if downstairs officials have spotted a circle (or an octagon), its precise location isn't as important as its existence near this part of the picture. A cat is a cat, even though its eyes were a bit closer to each other than those of an average kitty.

In addition to the House of Circles, one needs a *Triangular Ear House,* maybe also a *Whiskers Office* and a *Control Commission for Curly Tails.* There are tens of office blocks, some of them are taller and some process image data at a higher level. The combination of short lines into circles described above goes on at a higher level, where patches of pictures containing circles, crosses or triangles are assembled like puzzles into combinations of a few patches that form cat faces.

The final decision is made in the *Presidential Palace,* where a single number is combined out of all the location data that has been refined in complex ways. It attaches a "cat percentage" into the image that was fed into the system. If there's a cat in the picture, the right result is 100 percent. If there is no cat, the result is zero percent. The goal is to always get the right result, but we can be a bit flexible too. If the result is for example 88 and there's a cat in the picture, it's understandable: 88 is rounded up to 100 and not down to 0, after all.

Who has managed to draft such good instructions to this cat identification village that is full of office blocks? It has happened automatically during the time required by machine learning and through the use of mathematical optimization. Millions of cat pictures and a corresponding number of pictures without cats have been used for the teaching. The computational teaching work is incomprehensibly great for the human mind, and it consists of going systematically through the photos.

The offices have been handed photos one at a time. All bureaucrats work according to their instructions. The result has been checked, whether it predicts correctly the existence of a cat in the photo. If a cat picture gets less than a 100 percent or a picture without a cat more than zero, all instructions are modified a bit in the direction that's a bit closer to the right computational "cat percentage". The modifications are again done by optimizing. After millions of corrections, the neural net begins to identify the cat pictures correctly.

The huge thing about all of this is that none of the calculations is complex by itself. Adding and multiplying, choosing the biggest number and replacing negative numbers with a zero give us a computational machine that recognizes a cat out of a photo. That is without looking at a photo with eyes and without interpreting it with brains. Of course, kind of electric brain is running, because the action of each "bureaucrat" imitates the known action of a nerve cell in our brain.

This raises philosophical questions: Is the complexity of our brain just a matter of there being so many elements or nerve cells and so many connections between them? Could the increase of computing power lead to a situation where a computational method like the "office complex" I described could surpass our abilities in new fields of human endeavour, like love or art?

Modelling the Human Tastes for Art

All AI isn't based on the computational office bureaucracy I just described. It's possible to get astonishing results by using simpler computations.

Have you noticed, how good results the streaming services give you? I'm a friend of comedies and I often get very apt recommendations for the next show to view and to get me hooked on the movie. A mathematical model forms a probabilistic algorithm for my opinions based on what I've watched and how I've rated it. It would seem to magically know, how much I would like a movie or a series I haven't yet watched. How is it possible?

This is based on a mathematical technique called *compressed sensing.*

A streaming service has a million users and 10 million videos, for example. The users can attach a "thumbs up" or a "thumbs down" sign to tell the service firm and other users of their opinions. We can think of that information as a huge grid with a row for each user. Each TV series or movie has its own column on the other hand.

The streaming firm wishes to be in the following impossible situation: each work has a ready-made "thumbs up" or "thumbs down" sign as if the user had clicked them, even though he hasn't seen the work yet. Then it would be easy for the service to recommend films where the thumb points up and not the others.

But the new user doesn't have any markings yet. The few thumbs he adds to the system now and then concern material that he has already seen. How can we predict the grades a user gives to each video by using this very incomplete information?

Compressed methods of calculation have been developed during this millennium, and they come to the aid. They assume that there are only few explaining factors behind complex-looking phenomena. The compressed method seeks out these explaining factors by using the assumption that there are only a few of them. The explaining factor could be "Friends of romcoms like psychological thrillers too", but in practice they're more complex combinations of different styles.

Compressed sensing is an unbelievably powerful method. When the user has reviewed 25 videos or more, a mathematical model using compressed sensing can fill the right reviews for the whole material of 10 million films

(Candes & Plan, 2010) (https://wid.wisc.edu/tackling-netflix-problem-wid-researchers-snags-lagrange-prize/).

One gets the same kind of results in the field of music too. A mathematical formula called *principal content analysis* revealed that the musical tastes of test subjects was composed of these five elements:

1. Soft and relaxing music
2. An urban style with rhythmic genres like rap and jazz
3. Elegant music like classical, opera, "global music" and jazz.
4. Rough music that is loud and powerful
5. A down to earth, not very creative style of music, including e.g. country music.

Everybody's musical taste is some combination of the five. Everyone has his or her own emphasis, and undoubtedly a track can contain elements from more than one style. But the result is surprisingly simple: building an intermediate model of musical taste can be done with only five classifications (Rentfrow et al., 2011).

What does it tell about us human beings? At least I'd like to represent a magnificent cultural taste that can't be modelled with a simple formula. But it seemingly can. It looks like it's time to be humble before the force of mathematics.

5.3   Will Computers Replace Artists?

I earlier mentioned a successful Hemingway translation. It points to the direction that human translators won't be needed to translate literary texts in the future. I still feel that the idea is impossible, but mostly because I'd like to keep these kinds of high culture as human activities and outside the reach of machines. But the developments in recent years point in the direction that mathematical models that coarsely imitate the functions and interactions of our brain cells will replace human translators. Maybe experimental literature and poetry will be able to fight longer against the machines' ability to translate!

Translation is one thing, but the production of literary texts is another. Surely you can't replace writers with computers? You can trust computers with sports reports, investment news and other more schematic newspaper articles, we've seen that. But not creative literary writing? An AI is already a writer's companion too, at least in an assistant role. The American writer Robert Sloan has written an AI program that proposes endings for his sentences.

The *New York Times* visited to watch Sloan at work (Streitfeld, 2018). His novel takes place in future California, whose nature is recovering from hard times. "The bisons are back", he writes. "Herds 50 miles long." He doesn't finish the next sentence: "*The bison are gathered around the canyon ….*" What next? Sloan presses the tab key, and the computer beeps and starts thinking. Soon it proposes the continuation "by the bare sky." "Would I have written 'bare sky' by myself?" Sloan wonders. "Maybe, maybe not." And on: "*The bison have been traveling for two years back and forth ….*" A tab key, a beep and a proposed continuation for the text: "*between the main range of the city.*" Sloan praises: "That wasn't what I was thinking at all, but it's interesting. The lovely language just pops out and I go, 'Yes.'"

The sentences illustrate how a writer could use a neural network aid to creatively produce texts. What will this lead to in the future? Sloan muses like this: "I like this tool, although it's quite primitive. Next versions will make this look like a hundred years old crystal radio."

What about visual art? If a neural net can identify a cat from a picture, is it able to draw them too? After the neural network has been taught and when it copes well in its task of identifying cats, it can be run as if in reverse. The network can be put to produce pictures that obey as closely as possible the *First, Second* and *Third etc. instructions* of the virtual office blocks so that they produce as big numbers as possible for the bureaucrats to process. These pictures are as if they had been picked up from the LSD dreams of a 1960s hippie commune: they have eyes and other cat-like body parts here and there, blended in a background. You can easily see these pictures by running a web search with the keywords "deep dream cat image". The painter Salvador Dali could use this tool like Robin Sloan uses his text production bot!

These pictorial neural networks too are based on large picture archives, where they pick up characteristic features in their learning processes. It might be difficult to see pictures produced with this method as creative artistic work, although human artists too use pictures made by others as an inspiration.

In a way, neural networks are already replacing visual artists. They can modify a given picture to resemble the style of a given artist. There are a lot of available tools that are based on this idea, for example the cell phone apps *Prisma* and *Artisto*, and the DeepArt.io website. I was greatly astonished when I saw the results of these kinds of programs for the first time. I was in a scientific imaging conference in Albuquerque, New Mexico. The lecturer showed a photo of a canal in Amsterdam, and next the same photo styled in the way of Vincent van Gogh and then Piet Mondrian.

The neural network has passed a large number of van Gogh paintings through it, letting the hard-working bureaucrats in its office buildings send numbers to each other and fixing the *instructions* step by step so that they highlight van Gogh's way of handling outlines and coloured surfaces. So the set of instructions for mathematical operations has managed to learn the style of the great master and can modify other pictures to look like it.

These mathematical aids can be seen as tools that save the artist's time. For example, if a graphic artist looks for a certain atmosphere in an ad by using a pointillist expression, he doesn't have to dab 1000 of dots of oil colour on the canvas. He can photograph or draw the needed still life or landscape and use the method of neural network to add a pointillist style to it. After that extra edits take a lot less work.

It remains to be seen if a neural network AI can create a new visual style or even a new school of visual art someday. We aren't there yet, but the neural network learned to be the chess champion of the Universe in 4 h too.

References

Candes, E. J., & Plan, Y. (2010). Matrix completion with noise. *Proceedings of the IEEE, 98*(6), 925–936. https://doi.org/10.1109/JPROC.2009.2035722

Lewis-Kraus, G. (2016, December 14). The Great A.I. Awakening. *The New York Times Magazine.*

Rentfrow, P. J., Goldberg, L. R., & Levitin, D. J. (2011). The structure of musical preferences: A five-factor model. *Journal of Personality and Social Psychology, 100*(6), 1139–1157. https://doi.org/10.1037/a0022406

Silver, D., Hubert, T., Schrittwieser, J., Antonoglou, I., Lai, M., Guez, A., Lanctot, M., Sifre, L., Kumaran, D., Graepel, T., & Lillicrap, T. (2017). *Mastering chess and Shogi by self-play with a general reinforcement learning algorithm.*

Streitfeld, D. (2018, October 18). Computer stories: A.I. Is Beginning to Assist Novelists. *New York Times.*

6

Mathematics Belongs to Everyone

Many believe that mathematical skills are given at birth, if at all. But that's not true. Mathematical skill is born through practice, and everybody can become better at math.

Of course, some people have more mathematical tendencies than others. Let's compare mathematics with the skill of playing a guitar, which everyone can learn by going to guitar classes led by a good teacher at the local community college. I learnt too, thanks to Matti Wallenius! Still, teaching doesn't turn everyone to a guitar master like the musicians Erja Lyytinen or Alexi Laiho, but hard practice gives everyone skills to play the guitar to accompany singing songs together. I recommend that a beginner's goal is to learn *La Bamba.* Everybody can learn basic mathematical skills, just like they can learn basic guitar skills.

Everyone should study some mathematics, because math is a gym of the mind. Nobody needs a bench press or a jumping jack in everyday life, but they help when lifting a box when moving, or running to a bus. The consistency and sense of magnitudes that mathematics brings too helps when planning a mortgage or a list of shifts at work.

S. Siltanen, *Step into the World of Mathematics*,
https://doi.org/10.1007/978-3-030-73343-8_6

In our digital world, mathematics is used to run society and our everyday lives. In the past, one waved a bus ticket at the driver, and the ticket was made of cardboard. Now one waves a card full of digital data at a reader, and the reader picks up the data via radio frequencies. In the past, photography was chemistry, but now it's the mathematics of pixel grids. Patient records used to be written on paper in hospitals, but now they're fed into electronic computer systems. Mathematics has ended up everywhere in our lives via computer programs. Therefore we must have mathematicians just like we have electricians and painters.

Computer mathematics also opens up new possibilities for hobbyists. It works as the paintbrush of the artist, the sound effect of the musician, and the health data processor of the fitness fanatic. If you want to program a new hit game or write a fitness app, you need a mathematical toolbox.

The world of mathematics should be open to everyone. First, mathematics needs a more diverse crew to develop it from here on. Second, mathematical skills are like literacy: everyone has a right to them. Third, the needs and challenges of future mathematics are so complex and cross-disciplinary that mathematical knowledge is not enough to solve them. Mathematicians must be able to discuss with representatives of other disciplines, so that everyone can find suitable mathematical solutions to their problems.

That's why we need a bigger and more diverse group of people to do and develop mathematics. It's especially important to get girls into mathematics from an early age, so that a greater portion of them will end up as professional mathematicians when they grow up.

6.1 The Myth of Genius Is Destructive

The idea of my book is to invite everybody to the world of mathematics. But to be honest, I've got to tell you that something is wrong in this world.

Even though Finland is known as a beacon of equality in the world, this isn't the case in mathematics. I work in the Mathematics Department of the University of Helsinki. During its 379 years, it has hired only one woman as a professor: Sangitha Kulathinal started working as a professor of statistics in March 2019. The situation is not much better elsewhere in Finland. Less than 10 women have ever worked as math professors in Finland.

I can't draw any other conclusion than that the world of mathematics is unfair to women.

And this is not just a Finnish problem. Gender equality in mathematics is a bit better in the US, for example. However, the gender ratio is very biased towards men there too. Why? The phenomenon is complex and there are many factors influencing it. I'll take up one of them.

Alongside theoretical physics and philosophy, mathematics is a discipline where a *myth of genius* is widespread. According to the myth, only an exceptional individual can succeed in these endeavours. The individual has the following traits:

- He shows an astonishing ability at mathematics already in school, but is not interested in other subjects.
- As a researcher, he focuses on a narrow subject that is not understood by others. He wins the Fields Medal, the Abel Prize and the Nobel Prize, and holds an incomprehensible speech.

- He can't cope with everyday life or social situations. He suffers from mental health problems. He can't teach or supervise students.

I exaggerated a bit, but not much.

There are genius mathematicians corresponding to the myth, like the Nobel prize winner and schizophrenic John Nash, the codebreaker Alan Turing and the wunderkind Ramanujan, in whose dreams Hindu gods offered him amazing mathematical formulae. The movies *A Beautiful Mind, The Imitation Game* and *The Man who Knew Infinity* tell about them.

The list of geniuses also includes Andrew Wiles, who locked himself in a room for 1 years and proved Fermat's last theorem; Georg Cantor, who developed different infinities and spent a lot of time in mental hospitals; the St. Petersburg hermit Grigori Perelman, who solved the Poincare conjecture and refused the million dollar prize; and of course Évariste Galois, whose theory brought clarity to the problems of formulae for solving fifth- and higher-degree equations. Galois was killed in a duel when he was 19.

There are many fictional examples in entertainment media: Sheldon in *Big Bang Theory*, the janitor in *Good Will Hunting* and the autistic Raymond in *The Rain Man.*

The biggest problem with the genius myth is this: studies show that almost everyone has the subconscious assumption that the genius is male. I believe that this is one of the reasons, why the gender balance of the international mathematics community is badly biased. If the committee hiring a professor or a fellow has the hiring of a genius in mind, they have a subconscious bias against women applying for the job.

The situation is like auditions in classical music. The choices became more equal, when the musician was playing behind the curtain and only the music was audible.

Don't get me wrong on this genius matter. I greatly value the work of the researchers listed above. They made important scientific breakthroughs. I know many able mathematicians, whose technical skills I can only dream of and whose intellect I admire with awe. And not all are men! They're like music prodigies, who develop into virtuoso pianists or heavy rock guitarists through years of practice.

But mythical and narrow exceptional talent cannot be the only way of participating in mathematics. Science and technology need other kinds of practitioners too. I list a few alternatives to mythic geniuses below. Every one of them has a secret weapon, which the mythic genius doesn't have.

The Absorbing Teacher   There are many inspiring lecturers among professional mathematicians. They pull students into explorations and adventures in the world of mathematics. They know to challenge the youngsters just the right amount, without discouraging them with too difficult exercises. This allows the hatchlings to grow into new mathematicians for future needs. The secret weapons of an absorbing teacher are charisma and interpersonal skills.

The Generalist   This rare kind of researcher has studied many fields of mathematics, and hasn't necessarily gone very deep into any of them. They also know physics, computer science, philosophy, biology and linguistics. The secret weapon of the generalist: the ability to create new knowledge by combining completely different kinds of things.

The Science Politician   This suave social animal effortlessly builds international teams that power big congresses and scientific societies. They assess funding proposals in the EU and give expert advice to parliaments. They are among those who decide prize winners and formulate big science funding projects. The secret weapon of the science politician: skill at guiding scientific fads.

The Popularizer   Mathematics is a technical art that may look incomprehensible to an outsider. When popularizing it, one cannot use formulae or suppose that the hearer has deep knowledge of topology or functional analysis. Musicians don't introduce the listeners to notes or lecture them about counterpoints, they play music and sing. One must simplify the content but preserve the main points of mathematical ideas, when formulating them so that they are widely accessible. The secret weapon of the popularizer is the ability to tell about science in a way that puts oneself in the position of the hearer.

The Applied Modeller   Applying mathematics with intermediate models requires suitable simplifications, an understanding of computing and listening to the end user. The applied modeller knows physics and corporate product development. Their secret weapon is a service ethos.

As we see from this list, the mythical genius is not the only kind of researcher who can advance science. In fact, science requires diverse teams to develop, and it's therefore best to combine different types of researchers in a suitable mix.

One should think of the balance between different kinds of experts when hiring mathematicians. Selection boards should have a couple of different profiles in mind when hiring, like *the absorbing teacher, the generalist, the science politician, the popularizer* and *the applied modeller.*

The myth of genius is embedded deep into our mathematical culture. This objection arises always when thinking of other kinds of merits than the ones of the mythical genius: "If we do so, we lower the bar and don't hire the best applicant." But there's another way of looking at it. If we first choose the kind of researcher we use to assess the applicant, we can look for the highest quality *among researchers of that kind.*

Using many kinds of researchers in assessments has the extra advantage of reducing unconscious discrimination. It's better for science and fixes the gender bias in mathematics with a fairer one.

Next I'll take a look at the applied modeller type, because it's closest to the message of this books.

6.2 The Properties of an Applied Modeller

The applied modeller works in a service profession. She develops mathematical ideas that can be turned into computer programs and be of use for people looking to solve a given problem.

For example, a dentist may need a program that shows her the distance between nerves and the roots of a wisdom tooth with a smaller dose of X-rays. Or a developer of elevators or escalators may be looking for the easiest possible travel routes in a future shopping centre. A high school student could be looking for a filter on Instagram that matches his current mood.

All of these problems can be solved with mathematical models and computer programs.

There's a big difference between assessing the results of a theoretician and an applied modeller. In theoretical work, it's the *mathematics community* that decides whether a result is interesting or possibly even important. This arrangement is prone to inbreeding, especially because the results of abstract mathematics can usually be understood only by a small group of experts. Who decides which results of pure mathematics are good ones? Often a mathematician who is so famous that he has ended up in the role of a pathfinder. But how did he get there?

The situation is different in applied mathematics, because success depends on the satisfaction of end users. Is the dentist able to remove wisdom teeth safely with the new low-dose X-ray device? Are the customers able to move smoothly from floor to floor in the shopping centre, and does the elevator company get new orders with a successful example? Does the teenager dare to send a filtered selfie on the Internet? If not, the blame falls on the quality of the work of the applied mathematician.

The situation is similar in physics. There new theories are tested with measurements and assessed how well they describe natural phenomena. In physics, it is nature itself that is the outsider assessing the work, in applied mathematics it is the end users.

Therefore one of the most important task for the applied mathematician is to study closely the problems of the end user. It requires lengthy conversations and understanding the language and the mind-set of the other party.

In my career I have had a chance to study the basics of dentistry, how a nuclear plant functions, the formation of a human voice, the properties of counterfeit banknotes, welding metal pipes, making paper, clogging oil pipes, photography, diseases of blood circulation in the brain, cystic fibrosis, coronary artery disease, detecting breast cancer, plant metabolism and fossil jaws. I don't know a lot about any of these subjects, but still enough to build a reasonably accurate intermediate model of them.

I remember retirement talk by a Finnish mathematician. He said something like this when recalling his career: "Once we solved the problem in the two-dimensional case, we started to think about the right way to define it in higher dimensions. We could do it, but *it would be easy*. So we started doing this and proved ..." My ear picked up the phrase "it would be easy".

This is a big difference with applied mathematics: applied mathematics looks for the simplest solution to the problem of the end user. The situation is great if you can find one. An easy mathematical formula can be efficiently implemented with a computer, so calculations are quick. Unexpected problems are rarer with simple mathematical tools than with complex ones.

Usually in applied research projects difficulties pop up when dealing with real data, and a simple trick won't solve the problem. But the approach is the opposite as in pure mathematics: one does not look for hardships on purpose.

The situation is similar to space flight, where one needs solutions that are known to work. One doesn't install the newest microprocessor models to the NASA Mars Rovers, one installs electric components that have been tested for decades in difficult conditions instead. In space, the equipment must cope with terrible changes of temperature, shakes from the rocket engines and bumps from the landing, long-term radiation exposure and surprising electric blackouts.

The computational solutions designed for end users are the space probes of mathematics. When building them, one must similarly choose only approaches that can cope when the real world offers nasty surprises. There are unusual errors in measurements, tired users may feed odd inputs and the Internet connection could be cut at any time. One should not offer brand-new developments in mathematical theory for these conditions, because they are as sensitive as opera divas. One writes traditional, stable formulae for practical applications, like averages and carefully regularized computational inversion methods.

The applied mathematician isn't usually an expert of a narrow field. The problem described by the end user is sometimes best solved with a probability calculation, sometimes by cleverly filtering oscillating signals and often by using optimization. Therefore, it's good for the applied mathematician to be a bit of a jack of all trades. The store of his mathematical knowledge is like a buffet table: there are many kinds of things on the table to choose from, and none of the foods represents the latest trend from a top chef. Plain everyday products are enough to make up a workable and delicious dish.

6.3 A Challenge for Raising Enthusiasm: let's Bring Everybody to the World of Mathematics!

Mathematics has an awkward property: all knowledge is built on previous knowledge. Therefore one should practice math throughout the years at school, so that a break won't cause problems with learning new things in the future.

I'm raising a challenge to teachers, parents, educators, and people who use mathematics at work. I promise to tell kids and youngsters about mathematics in as excitingly as I can, to encourage them to trust their ability to learn mathematics and to remind them to do their homework. I also promise to change the structures of the scientific world so that they are fairer so that a more equal and diverse group of people will work as mathematicians in the future.

You can respond to the challenge simply like this: encourage every child and adult in their mathematical studies and efforts. And if you think that some people are innately untalented in mathematics, try to get rid of that idea.

Printed in the United States
by Baker & Taylor Publisher Services